Noise Sensitivity of Boolean Functions and Percolation

This is a graduate-level introduction to the theory of Boolean functions, an exciting area lying on the border of probability theory, discrete mathematics, analysis, and theoretical computer science. Certain functions are highly sensitive to noise; this can be seen via Fourier analysis on the hypercube. The key model analyzed in depth is critical percolation on the hexagonal lattice. For this model, the critical exponents, previously determined using the now-famous Schramm–Loewner evolution, appear here in the study of sensitivity behavior. Even for this relatively simple model, beyond the Fourier-analytic setup, there are three crucially important but distinct approaches: hypercontractivity of operators, connections to randomized algorithms, and viewing the spectrum as a random Cantor set. This book assumes a basic background in probability theory and integration theory. Each chapter ends with exercises, some straightforward, some challenging.

CHRISTOPHE GARBAN is a professor of mathematics at University of Lyon 1, France.

JEFFREY E. STEIF is a professor of mathematical sciences at Chalmers University of Technology in Gothenburg, Sweden.

INSTITUTE OF MATHEMATICAL STATISTICS
TEXTBOOKS

IMS Textbooks give introductory accounts of topics of current concern suitable for advanced courses at master's level, for doctoral students, and for individual study. They are typically shorter than a fully developed textbook, often arising from material created for a topical course. Lengths of 100–290 pages are envisaged. The books typically contain exercises.

Other Books in the Series

1. *Probability on Graphs*, by Geoffrey Grimmett
2. *Stochastic Networks*, by Frank Kelly and Elena Yudovina
3. *Bayesian Filtering and Smoothing*, by Simo Särkkä
4. *The Surprising Mathematics of Longest Increasing Subsequences*, by Dan Romik

Noise Sensitivity of Boolean Functions and Percolation

CHRISTOPHE GARBAN
ICJ, Université Lyon 1, Lyon

JEFFREY E. STEIF
Chalmers University of Technology, Gothenburg

CAMBRIDGE
UNIVERSITY PRESS

CAMBRIDGE
UNIVERSITY PRESS

Shaftesbury Road, Cambridge CB2 8EA, United Kingdom

One Liberty Plaza, 20th Floor, New York, NY 10006, USA

477 Williamstown Road, Port Melbourne, VIC 3207, Australia

314–321, 3rd Floor, Plot 3, Splendor Forum, Jasola District Centre, New Delhi – 110025, India

103 Penang Road, #05–06/07, Visioncrest Commercial, Singapore 238467

Cambridge University Press is part of Cambridge University Press & Assessment,
a department of the University of Cambridge.

We share the University's mission to contribute to society through the pursuit of
education, learning and research at the highest international levels of excellence.

www.cambridge.org
Information on this title: www.cambridge.org/9781107076433

© Christophe Garban and Jeffrey E. Steif 2015

First published 2015

A catalogue record for this publication is available from the British Library

Library of Congress Cataloging-in-Publication data
Garban, Christophe, 1982–
Noise sensitivity of Boolean functions and percolation / Christophe Garban, Université
Lyon 1, Lyon, Jeffrey E. Steif, Chalmers University of Technology, Gothenburg.
pages   cm. – (Institute of Mathematical Statistics textbooks)
Includes bibliographical references and index.
ISBN 978-1-107-07643-3 (hardback) –
ISBN 978-1-107-43255-0 (pbk)
1. Statistical physics–Textbooks.   2. Percolation (Statistical physics)–Textbooks.
3. Algebra, Boolean–Textbooks.   I. Steif, Jeffrey E.   II. Title.
QC174.8.G36   2015
530.13–dc23   2014023801

ISBN   978-1-107-07643-3   Hardback
ISBN   978-1-107-43255-0   Paperback

To Laure, Victor, Camille,
Aila, Adam, and Noah

Contents

Preface

The purpose of this book is to present a self-contained theory of Boolean functions through the prism of statistical physics. The material presented here was initially designed as a set of lecture notes for the 2010 Clay summer school, and we decided to maintain the informal style which, we hope, will make this book more reader friendly.

Before going into Chapter 1, where precise definitions and statements are given, we wish to describe in an informal manner what this book is about. Our main companion through the whole book will be what one calls a **Boolean function**. This is simply a function of the following type:[1]

$$f : \{0, 1\}^n \to \{0, 1\}.$$

Traditionally, the study of Boolean functions arises more naturally in theoretical computer science and combinatorics. In fact, over the last 20 years, mainly thanks to the computer science community, a very rich structure has emerged concerning the properties of Boolean functions. The first part of this book (Chapters 1 to 5) is devoted to a description of some of the main achievements in this field. For example, a crucial result that has inspired much of the work presented here is the so-called KKL theorem (for Kahn–Kalai–Linial, 1989), which in essence says that any "reasonable" Boolean function has at least one variable that has a large influence on the outcome (namely at least $\Omega(\log n/n)$). See Theorem 1.14.

The second part of this book is devoted to the powerful use of Boolean functions in the context of statistical physics and in particular in percolation theory. It was recognized long ago that some of the striking properties that hold in great generality for Boolean functions have deep implications in statistical physics. For example, a version of the KKL theorem enables one to recover in an elegant manner the celebrated theorem of Kesten from

---

[1] In fact, in this book we view Boolean functions rather as functions from
$\{-1, 1\}^n \to \{-1, 1\}$ because their Fourier decomposition is then simpler to write down; nevertheless this is still the same combinatorial object.

1980 that states that the critical point for percolation on $\mathbb{Z}^2$, $p_c(\mathbb{Z}^2)$, is $1/2$. More recently, Beffara and Duminil-Copin used an extension of this KKL property obtained by Graham and Grimmett to prove the conjecture that $p_c(q) = \frac{\sqrt{q}}{1+\sqrt{q}}$ for the Fortuin–Kasteleyn percolation model with parameter $q \geq 1$ in $\mathbb{Z}^2$. It is thus a remarkable fact that general principles such as the KKL property are powerful enough to capture some (not all) of the main technical difficulties that arise in understanding the phase transitions of various statistical physics models.

In the 1990s, Talagrand as well as Benjamini, Kalai, and Schramm pushed this connection between Boolean functions and statistical physics even further. In 1998, Benjamini, Kalai, and Schramm introduced the fruitful concept of **noise sensitivity** of Boolean functions. Their main motivation was to study the behavior of critical percolation, but let us briefly explain what noise sensitivity corresponds to in the more common situation of *voting schemes*. Suppose n voters have to decide between two candidates denoted by 0 and 1. They first have to agree on a voting procedure or *voting scheme*, which may be represented by a Boolean function $f: \{0,1\}^n \to \{0,1\}$. In France or Sweden, this Boolean function would simply be the majority function on n bits whereas in the United States, the Boolean function f would be more complicated: in fancy words, it would correspond to an iterated weighted majority function on $n \approx 10^8$ voters. The collection of all votes is a certain configuration ω in the hypercube $\{0,1\}^n$. If the election is "close," it is reasonable to consider $\omega = (x_1, \ldots, x_n)$ as a uniform point chosen in $\{0,1\}^n$. In other words, we assume that each voter $i \in [n]$ tosses a fair coin in $\{0,1\}$ and votes accordingly. As such the *true* result of the election should be the output $f(\omega) = f(x_1, \ldots, x_n) \in \{0,1\}$. In reality the *actual* result of the election will rather be the output $f(\omega_\epsilon)$, where ω is an ϵ-perturbation of the configuration ω. Roughly speaking, we assume that independently for each voter $i \in [n]$, an error occurs (meaning that the value of the bit is flipped) with probability a fixed parameter $\epsilon > 0$. See Chapter 1 for precise definitions. In this language, a **noise-sensitive Boolean function** is a function for which the outputs $f(\omega)$ and $f(\omega_\epsilon)$ are almost independent of each other even with a very small level of *noise* ϵ. As an example, the *parity* function defined by $f(x_1, \ldots, x_n) = 1_{\sum x_i \equiv 1 \bmod 2}$ is noise sensitive as $n \to \infty$.

As we will see, there is a very useful spectral characterization of noise sensitivity. Indeed, in the same way as a real function $g: \mathbb{R}\backslash\mathbb{Z} \to \mathbb{R}$ can be decomposed into $g(x) = \sum_{n \in \mathbb{Z}} \hat{g}(n)e^{2i\pi nx}$, one can decompose a Boolean function $f: \{0,1\}^n \to \{0,1\}$ into a Fourier–Walsh series $f(\omega) = \sum_{S \subseteq [n]} \hat{f}(S)\chi_S(\omega)$. (See Chapter 4 for details.) Noise-sensitive

Boolean functions are precisely the Boolean functions whose spectrum is concentrated on "high frequencies"; that is, most of their Fourier coefficients (in a certain analytic and quantitative manner) correspond to subsets $S \subseteq [n]$ with $|S| \gg 1$.

It is thus not surprising that with such a spectral characterization of noise sensitivity, a significant part of this book is devoted to various techniques that allow us to detect high-frequency behavior for Boolean functions. The techniques we introduce are essentially of three different flavors:

1. Analytical techniques based on **hypercontractive** estimates (Chapter 5)
2. A criterion based on randomized algorithms (Chapter 8)
3. A study of the "fractal" behavior of frequencies $S \subseteq [n]$ (Chapter 10)

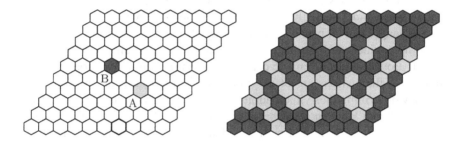

To make the link with statistical physics, consider the following Boolean function, which is well known in computer science and in game theory because it represents the solution of the **Hex game**. In the figure above, we represent a Hex game on a 10×10 table: player A tries to go from the left boundary to the right using gray hexagon tiles, while player B tries to go from the top boundary to the bottom using black hexagon tiles. The players either take turns or, as in the **random turn hex game**, they toss a coin at each turn to decide who will move. At the end of the game, we obtain a tiling of the table as in the figure and the result of the game is then described by a certain Boolean function $f_{10} : \{A, B\}^{100} \to \{A, B\}$. Note that in the figure, player A has won. As we will see, this Boolean function (or rather the family $\{f_n\}$ defined analogously on $n \times n$ tables) is instrumental in our study of how the model of percolation responds to small random perturbations.

Boolean functions of this type are notoriously hard to study, and this book develops tools aimed at understanding such Boolean functions. In particular, we will eventually see that as $n \to \infty$, most of the Fourier transform $\hat{f}_n$ of the Hex-function f_n on an $n \times n$ table is concentrated on frequencies of size $|S|$ about $n^{3/4}$. The appearance of the surprising exponent of $3/4$ corresponds to one of the **critical exponents** that are aimed at describing the fractal geometry of critical percolation. See Chapter 2.

This high-frequency behavior of the Hex-functions $\{f_n\}$ implies readily that critical planar percolation on the triangular lattice is highly sensitive to noise (in a quantitative manner given by the above $n^{3/4}$ asymptotics). This noise sensitivity of percolation has surprising consequences concerning the model of **dynamical percolation** (see Chapter 11). We will see among other things that there exist exceptional times at which an infinite *primal* cluster coexists with an infinite *dual* cluster (Theorem 11.9), which is a very counterintuitive phenomenon in percolation theory.

In Chapter 7, we give another application to statistical physics of a very different flavor: consider the random metric R on the lattice $\mathbb{Z}^d$, $d \geq 2$, where each edge is independently declared to be of length a with probability $1/2$ and of length b with probability $1/2$ (where $0 < a < b < \infty$ are fixed). Fascinating conjectures have been made about the fluctuations of the random R-ball around its deterministic convex limit. In particular, it is conjectured that these fluctuations are of magnitude $R^{1/3}$ in two dimensions and that the law describing these fluctuations is intimately related to the celebrated Tracy–Widom law that describes the fluctuations of the largest eigenvalue of large random Hermitian matrices. This book certainly does not settle this stunning conjecture but it does present the best results to date on the fluctuations of this metric using a Fourier approach.

This book is structured as follows: in Chapters 1, 3, 4, 5, 8, and 9, we introduce general tools for Boolean functions that are applicable in various settings (and are thus not restricted to the context of statistical physics). Several examples in these chapters illustrate links to other active fields of mathematics. Chapter 2 is a short introduction to the model of percolation. Chapters 6, 10, and 11 are more specifically targeted toward the analysis of the **noise sensitivity of critical percolation** as well as its consequences for dynamical percolation. Chapter 7 analyzes the fluctuations of the earlier mentioned random metrics on $\mathbb{Z}^d$, $d \geq 2$. Chapter 12 explores a large variety of interesting topics tangential to the main contents of this book. Finally Chapter 13 collects some open problems.

We assume readers have the mathematical maturity of a first-year graduate student and a reasonable background in probability theory and integration theory. Having seen some percolation would be helpful but not neccesary.

Acknowledgments

This book would not be in its present shape without the help of many people whom we now wish to warmly thank. First of all, we thank Charles Newman, Vladas Sidoravicius, and Wendelin Werner for inviting us to give a course on this topic at the 2010 Clay summer school in Buzios. (This resulted in a set of lecture notes published by the Clay Institute that was the starting form of the present book.) This book would not exist without having had the opportunity to give such a course.

We also thank Diana Gillooly for her constant help at Cambridge University Press.

Many people helped us greatly with their remarks, which improved the manuscript. In this respect, we wish in particular to thank Ragnar Freij, who served as a very good teaching assistant for this course at the Clay Institute. We also thank Juhan Aru, Vincent Beffara, Itai Benjamini, Ehud Friedgut, Johan Håstad, Jeff Kahn, Gil Kalai, Jean-Christophe Mourrat, Ryan O'Donnell, Malin Palö, Yuval Peres, Gábor Pete, Jeremy Quastel, Mike Saks, Mikaël de la Salle, Yilun Shang, Zhan Shi, Arvind Singh, Vincent Tassion, and Xiaolin Zeng.

Finally, at the time of concluding the writing of this book, our thoughts go out to Oded Schramm. This book would simply not exist had it not been for Oded.

Lyon and Gothenburg, October 1, 2014

CHRISTOPHE GARBAN and JEFFREY E. STEIF

Notations

Ω_n	hypercube $\{-1, 1\}^n$
$\mathbf{I}_k(f)$	influence of the kth variable on f
$\mathbf{I}_k^p(f)$	influence of the kth variable on f at level p
$\mathbf{I}(f)$	total influence of the function f; see Definition 1.10
$\mathbf{Inf}(f)$	influence vector of f
$\mathbf{H}(f)$	sum of the squared influences; see Definition 5.5
$\alpha_1(R)$	probability in critical percolation to have an open path from 0 to $\partial B(0, R)$
$\alpha_1(r, R)$	multiscale version of the above
$\alpha_4(R)$	probability of a *four-arm event* from 0 to $\partial B(0, R)$
$\alpha_4(r, R)$	multiscale version of the above
χ_S	character $\chi_S(x_1, \ldots, x_n) := \prod_{i \in S} x_i$
$\widehat{f}(S)$	Fourier coefficient $\widehat{f}(S) = \langle f, \chi_S \rangle = \mathbb{E}[f \chi_S]$
$E_f(m), 1 \le m \le n$	*energy spectrum* of f; see Definition 4.1
$\nabla_k f$	discrete derivative along k: $\nabla_k f(\omega) := f(\omega) - f(\sigma_k(\omega))$
$\mathcal{P} = \mathcal{P}(f)$	*pivotal set* of f; see Definition 1.7
$\mathscr{S} = \mathscr{S}_f$	*spectral sample* of f; see Definition 9.1
$\widehat{\mathbb{Q}}_f$	*spectral measure* of f; see Definition 9.1
$\widehat{\mathbb{P}}_f$	*spectral probability measure* of f; see Definition 9.2
$f(n) \asymp g(n)$	there exists some constant $C < \infty$ such that $C^{-1} \le \frac{f(n)}{g(n)} \le C$, $\forall n \ge 1$
$f(n) \le O(g(n))$	there exists some constant $C < \infty$ such that $f(n) \le C g(n)$, $\forall n \ge 1$
$f(n) \ge \Omega(g(n))$	there exists some constant $C > 0$ such that $f(n) \ge C g(n)$, $\forall n \ge 1$
$f(n) = o(g(n))$	$\lim_{n \to \infty} \frac{f(n)}{g(n)} = 0$

1

Boolean functions and key concepts

In this first chapter, we set the stage for the book by presenting many of its key concepts of the book and stating a number of important theorems that we prove here.

1.1 Boolean functions

Definition 1.1 A **Boolean function** is a function from the hypercube $\Omega_n := \{-1, 1\}^n$ into either $\{-1, 1\}$ or $\{0, 1\}$.

Ω_n is endowed with the uniform measure $\mathbb{P} = \mathbb{P}^n = (\frac{1}{2}\delta_{-1} + \frac{1}{2}\delta_1)^{\otimes n}$ and $\mathbb{E}$ denotes the corresponding expectation. Occasionally, Ω_n will be endowed with the general product measure $\mathbb{P}_p = \mathbb{P}_p^n = ((1-p)\delta_{-1} + p\delta_1)^{\otimes n}$ but in such cases the p is made explicit. $\mathbb{E}_p$ then denotes the corresponding expectation.

An element of Ω_n is denoted by either ω or ω_n and its n bits by $x_1, \ldots, x_n$ so that $\omega = (x_1, \ldots, x_n)$.

For the range, we choose to work with $\{-1, 1\}$ in some contexts and $\{0, 1\}$ in others, and at some specific places we even relax the Boolean constraint (i.e., that the function takes only two possible values). In these cases (which are clearly identified), we consider instead real-valued functions $f : \Omega_n \to \mathbb{R}$.

A Boolean function f is canonically identified with a subset A_f of Ω_n via $A_f := \{\omega : f(\omega) = 1\}$.

Remark Often, Boolean functions are defined on $\{0, 1\}^n$ rather than $\Omega_n = \{-1, 1\}^n$. This does not make any fundamental difference but, as we see later, the choice of $\{-1, 1\}^n$ turns out to be more convenient when one wishes to apply Fourier analysis on the hypercube.

1.2 Some examples

We begin with a few examples of Boolean functions. Others appear throughout this chapter.

Example 1.2 (Dictator)

$$\mathbf{DICT}_n(x_1,\ldots,x_n) := x_1.$$

The first bit determines what the outcome is.

Example 1.3 (Parity)

$$\mathbf{PAR}_n(x_1,\ldots,x_n) := \prod_{i=1}^{n} x_i.$$

This Boolean function's output is determined by whether the number of -1's in ω is even or odd.

These two examples are in some sense trivial, but they are good to keep in mind because in many cases they turn out to be the "extreme cases" for properties concerning Boolean functions.

The next rather simple Boolean function is of interest in social choice theory.

Example 1.4 (Majority function)   Let n be odd and define

$$\mathbf{MAJ}_n(x_1,\ldots,x_n) := \mathrm{sign}(\sum_{i=1}^{n} x_i).$$

Following are two further examples that also arise in our discussions.

Example 1.5 (Iterated 3-Majority function)   Let $n = 3^k$ for some integer k. The bits are indexed by the leaves of a rooted 3-ary tree (so the root has degree 3, the leaves have degree 1, and all others have degree 4) with depth k. Apply Example 1.4 (with $n = 3$) iteratively to obtain values at the vertices at level $k-1$, then level $k-2$, and so on until the root is assigned a value. The root's value is then the output of f. For example, when $k = 2$, $f(-1,1,1;1,-1,-1;-1,1,-1) = -1$. The recursive structure of this Boolean function enables explicit computations for various properties of interest.

Example 1.6 (Clique containment)   If $r = \binom{n}{2}$ for some integer n, then Ω_r can be identified with the set of labeled graphs on n vertices. (Bit x_i is 1 if and only if the ith edge is present.) Recall that a **clique** of size k of a graph $G = (V, E)$ is a complete graph on k vertices embedded in G.

Now for any $1 \leq k \leq \binom{n}{2} = r$, let **CLIQ**$_n^k$ be the indicator function of the event that the random graph G_ω defined by $\omega \in \Omega_r$ contains a clique of size k. Choosing $k = k_n$ so that this Boolean function is nondegenerate turns out to be a rather delicate issue. The interesting regime is near $k_n \approx 2\log_2(n)$. See Exercise 1.9 for this "tuning" of $k = k_n$. It turns out that for most values of n, the Boolean function **CLIQ**$_n^k$ is degenerate (i.e., has small variance) for all values of k. However, there is a sequence of n for which there is some $k = k_n$ for which **CLIQ**$_n^k$ is nondegenerate.

1.3 Pivotality and influence

This section contains our first fundamental concepts. We abbreviate $\{1, \dots, n\}$ as $[n]$.

Definition 1.7 Given a Boolean function f from Ω_n into either $\{-1, 1\}$ or $\{0, 1\}$ and a variable $i \in [n]$, we say that i **is pivotal for** f for ω if $f(\omega) \neq f(\omega^i)$ where ω^i is ω but flipped in the ith coordinate. Note that this event $\{f(\omega) \neq f(\omega^i)\}$ is measurable with respect to $\{x_j\}_{j \neq i}$.

Definition 1.8 The **pivotal set**, $\mathcal{P}$, for f is the random set of $[n]$ given by

$$\mathcal{P}(\omega) = \mathcal{P}_f(\omega) := \{i \in [n] : i \text{ is pivotal for } f \text{ for } \omega\}.$$

Expressed in words, the pivotal set is the (random) set of bits with the property that if you flip the bit, then the function output changes.

Definition 1.9 The **influence** of the ith bit, $\mathbf{I}_i(f)$, is defined by

$$\mathbf{I}_i(f) := \mathbb{P}(\, i \text{ is pivotal for } f \,) = \mathbb{P}(f(\omega) \neq f(\omega^i)) = \mathbb{P}(i \in \mathcal{P})$$

Let also the **influence vector**, $\mathbf{Inf}(f)$, be the collection of all the influences: i.e. $\{\mathbf{I}_i(f)\}_{i \in [n]}$.

Expressed in words, the influence of the ith bit is the probability that, on flipping this bit, the function output changes. This concept originally arose in political science to measure the power of different voters and is often called the Banzhaf power index (see (B65)) but in fact the concept arose earlier (see (P46)) in the work of L. Penrose.

Definition 1.10 The **total influence**, $\mathbf{I}(f)$, is defined by

$$\mathbf{I}(f) := \sum_i \mathbf{I}_i(f) = \|\mathbf{Inf}(f)\|_1 = \mathbb{E}(|\mathcal{P}|).$$

It would now be instructive to compute these quantities for Examples 1.2–1.4. See Exercise 1.1.

Later, we will need the last two concepts in the context where our probability measure is $\mathbb{P}_p$ instead. We now give the corresponding definitions.

Definition 1.11   The **influence vector at level** p, $\{\mathbf{I}_i^p(f)\}_{i\in[n]}$, is defined by

$$\mathbf{I}_i^p(f) := \mathbb{P}_p(\, i \text{ is pivotal for } f\,) = \mathbb{P}_p(f(\omega) \neq f(\omega^i)) = \mathbb{P}_p(i \in \mathcal{P}).$$

Definition 1.12   The **total influence at level** p, $\mathbf{I}^p(f)$, is defined by

$$\mathbf{I}^p(f) := \sum_i \mathbf{I}_i^p(f) = \mathbb{E}_p(|\mathcal{P}|).$$

It turns out that the total influence has a geometric-combinatorial interpretation as the size of the so-called edge boundary of the corresponding subset of the hypercube. See Exercise 1.4.

Remark Aside from its natural definition as well as its geometric interpretation as measuring the edge boundary of the corresponding subset of the hypercube, the notion of *total influence* arises very naturally when one studies **sharp thresholds** for *monotone functions* (to be defined in Chapter 3). Roughly speaking, as we see in detail in Chapter 3, for a monotone event A, $d\mathbb{P}_p[A]/dp$ is the total influence at level p (this is the Margulis–Russo formula). This tells us that the speed at which things change from the event A "almost surely" *not* occurring to the case where it "almost surely" *does* occur is very sudden if the Boolean function happens to have a large total influence.

1.4 The Kahn–Kalai–Linial theorems

This section addresses the following question. Does there always exist some variable i with (reasonably) large influence? In other words, for large n, what is the smallest value (as we vary over Boolean functions) that the largest influence (as we vary over the different variables) can take on?

Because for the constant function all influences are 0, and the function that is 1 only if all the bits are 1 has all influences $1/2^{n-1}$, clearly we want to deal with functions that are reasonably balanced (meaning having variances not so close to 0) or, alternatively, obtain lower bounds on the maximal influence in terms of the variance of the Boolean function.

The first result in this direction is the following. A sketch of the proof is given in Exercise 1.5.

Theorem 1.13 (Discrete Poincaré) *If f is a Boolean function mapping Ω_n into $\{-1, 1\}$, then*

$$\mathrm{Var}(f) \le \sum_i \mathbf{I}_i(f).$$

It follows that there exists some i such that

$$\mathbf{I}_i(f) \ge \mathrm{Var}(f)/n.$$

This gives a first answer to our question. For reasonably balanced functions, there is some variable whose influence is at least of order $1/n$. *Can we find a better "universal" lower bound on the maximal influence?* Note that for Example 1.4 all the influences are of order $1/\sqrt{n}$ (and the variance is 1). Therefore, in terms of our question, the universal lower bound we are looking for should lie somewhere between $1/n$ and $1/\sqrt{n}$. The following celebrated result improves by a logarithmic factor on the $\Omega(1/n)$ lower bound.

Theorem 1.14 (KKL88) *There exists a universal $c > 0$ such that if f is a Boolean function mapping Ω_n into $\{0, 1\}$, then there exists some i such that*

$$\mathbf{I}_i(f) \ge c\mathrm{Var}(f)(\log n)/n.$$

What is remarkable about this theorem is that this "logarithmic" lower bound on the maximal influence turns out to be *sharp*! This is shown by the following example by Ben-Or and Linial.

Example 1.15 (Tribes) Partition $[n]$ into disjoint blocks of length $\log_2(n) - \log_2(\log_2(n))$ with perhaps some leftover debris. Define f_n to be 1 if there exists at least one block that contains all 1's, and 0 otherwise.

One can check that the sequence of variances stays bounded away from 0 and that all the influences (including of course those belonging to the debris which are equal to 0) are smaller than $c(\log n)/n$ for some $c < \infty$. See Exercise 1.3. Hence Theorem 1.14 is indeed sharp. We mention that in (BOL87), the example of Tribes was given and the question of whether $\log n/n$ was sharp was asked.

Our next result tells us that if all the influences are "small," then the total influence is large.

Theorem 1.16 (KKL88) *There exists $c > 0$ such that if f is a Boolean function mapping Ω_n into $\{0, 1\}$ and $\delta := \max_i \mathbf{I}_i(f)$, then*

$$\mathbf{I}(f) \ge c\,\mathrm{Var}(f)\log(1/\delta).$$

Or equivalently,

$$\|\mathbf{Inf}(f)\|_1 \geq c \operatorname{Var}(f) \log \frac{1}{\|\mathbf{Inf}(f)\|_\infty}.$$

One can in fact talk about the influence of a set of variables rather than the influence of a single variable.

Definition 1.17 Given $S \subseteq [n]$, the **influence of** S, $\mathbf{I}_S(f)$, is defined by

$$\mathbf{I}_S(f) := \mathbb{P}(\ f \text{ is not determined by the bits in } S^c).$$

It is easy to see that when S is a single bit, this corresponds to our previous definition. The following is also proved in (KKL88). We do not give the proof in this book.

Theorem 1.18 (KKL88) *Given a sequence f_n of Boolean functions mapping Ω_n into $\{0, 1\}$ such that $0 < \inf_n \mathbb{E}_n(f) \leq \sup_n \mathbb{E}_n(f) < 1$ and any sequence a_n going to ∞ arbitrarily slowly, then there exists a sequence of sets $S_n \subseteq [n]$ such that $|S_n| \leq a_n n / \log n$ and $\mathbf{I}_{S_n}(f_n) \to 1$ as $n \to \infty$.*

Theorems 1.14 and 1.16 are proved in Chapter 5.

1.5 Noise sensitivity and noise stability

This section introduces our second set of fundamental concepts.

Let ω be uniformly chosen from Ω_n and let ω_ϵ be ω but with each bit independently "re-randomized" with probability ϵ. To rerandomize a bit means that, independently of everything else, the value of the bit is rechosen to be 1 or -1, each with probability $1/2$. Note that ω_ϵ then has the same distribution as ω.

The following definition is *central*. Let m_n be an increasing sequence of integers and let $f_n : \Omega_{m_n} \to \{\pm 1\}$ or $\{0, 1\}$.

Definition 1.19 The sequence $\{f_n\}$ is **noise sensitive** if for every $\epsilon > 0$,

$$\lim_{n \to \infty} \mathbb{E}[f_n(\omega) f_n(\omega_\epsilon)] - \mathbb{E}[f_n(\omega)]^2 = 0. \tag{1.1}$$

Because f_n takes just two values, this definition says that the random variables $f_n(\omega)$ and $f_n(\omega_\epsilon)$ are asymptotically independent for $\epsilon > 0$ fixed and n large. We see later that (1.1) holds for one value of $\epsilon \in (0, 1)$ if and only if it holds for all such ϵ. The following notion captures the opposite situation, where the two events are close to being the same event if ϵ is small, uniformly in n.

Definition 1.20   The sequence $\{f_n\}$ is **noise stable** if

$$\limsup_{\epsilon \to 0} \sup_n \mathbb{P}(f_n(\omega) \neq f_n(\omega_\epsilon)) = 0.$$

It is an easy exercise to check that a sequence $\{f_n\}$ is both noise sensitive and noise stable if and only if it is degenerate in the sense that the sequence of variances $\{\mathrm{Var}(f_n)\}$ goes to 0. Note also that a sequence of Boolean functions could be neither noise sensitive nor noise stable (see Exercise 1.11).

It is also an easy exercise to check that Example 1.2 (Dictator) is noise stable and Example 1.3 (Parity) is noise sensitive. We see later, when Fourier analysis is brought into the picture, that these examples are the two opposite extreme cases. For the other examples, it turns out that Example 1.4 (Majority) is noise stable, while Examples 1.5,1.6, and 1.15 are all noise sensitive. See Exercises 1.6–1.9. In fact, there is a deep theorem (see (MOO10)) that says in some sense that, among all low-influence Boolean functions, Example 1.4 (Majority) is the most stable.

In Figure 1.1, we give a slightly *impressionistic* view of what "noise sensitivity" is.

1.6 The Benjamini–Kalai–Schramm noise sensitivity theorem

We now come to the main theorem concerning noise sensitivity.

Theorem 1.21 (BKS99)   *If*

$$\lim_n \sum_k \mathbf{I}_k(f_n)^2 = 0, \tag{1.2}$$

then $\{f_n\}$ is noise sensitive.

Remark The converse of Theorem 1.21 is clearly false, as shown by Example 1.3. However, the converse is true for **monotone functions** (defined in the next chapter), as we see in Chapter 4.

Theorem 1.21 allows us to conclude noise sensitivity of many of the examples introduced in this first chapter. See Exercise 1.10. This theorem is proved in Chapter 5.

1.7 Percolation crossings: Our final and most important example

We have saved our most important example to the end. This book would not have been written were it not for this example and for the results that have been proved about it.

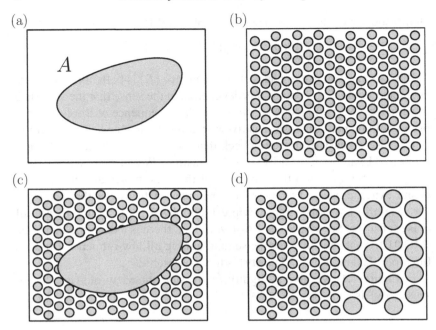

Figure 1.1 Consider the following "experiment": take a bounded domain in the plane, say a rectangle, and consider a measurable subset A of this domain. What would be an analog of the definitions of *noise sensitive* or *noise stable* in this case? Start by sampling a point x uniformly in the domain according to Lebesgue measure. Then apply some noise to this position x to end up with a new position x_ϵ. One can think of many natural "noising" procedures here. For example, let x_ϵ be a uniform point in the ball of radius ϵ around x, conditioned to remain in the domain. (This is not quite perfect as this procedure does not exactly preserve Lebesgue measure, but don't worry about this.) The natural analog of Definitions 1.19 and 1.20 is to ask whether $1_A(x)$ and $1_A(x_\epsilon)$ are decorrelated or not.

Question: According to this analogy, what are the sensitivity and stability properties of the sets A sketched in pictures (a) to (d)? Note that to match with Definitions 1.19 and 1.20, one should consider sequences of subsets $\{A_n\}$ instead, as noise sensitivity is an asymptotic notion.

Consider percolation on $\mathbb{Z}^2$ at the critical point $p_c(\mathbb{Z}^2) = 1/2$. (See Chapter 2 for a brief review of the model.) At this critical point, there is no infinite cluster, but somehow clusters are "large" and there are clusters at all scales. This can be seen using duality or with the RSW Theorem 2.1. To understand the geometry of the critical picture, the following large-scale *observables* turn out to be very useful: Let Ω be a piecewise smooth domain with two disjoint open arcs ∂_1 and ∂_2 on its boundary $\partial\Omega$. For each $n \geq 1$, we consider the scaled domain $n\Omega$. Let A_n be the event that there is an open path in ω from $n\partial_1$ to $n\partial_2$ which stays inside $n\Omega$. Such events are called **crossing events**. They are naturally associated with Boolean functions whose entries are indexed by the set of edges inside $n\Omega$ (there are $O(n^2)$ such variables).

For simplicity, consider the particular case of rectangle crossings.

Example 1.22 (Percolation crossings)

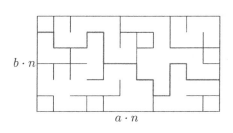

$b \cdot n$

$a \cdot n$

Let $a, b > 0$ and let us consider the rectangle $[0, a \cdot n] \times [0, b \cdot n]$. The left-to-right crossing event corresponds to the Boolean function $f_n: \{-1,1\}^{O(1)n^2} \to \{0,1\}$ defined as follows:

$$f_n(\omega): = \begin{cases} 1 & \text{if there is a left--} \\ & \text{right crossing} \\ 0 & \text{otherwise} \end{cases}$$

We later prove that this sequence of Boolean functions $\{f_n\}$ is noise sensitive. This means (see Exercise 4.7) that if a percolation configuration $\omega \sim \mathbb{P}_{p_c=1/2}$ is given, one typically cannot predict anything about the large-scale clusters of the slightly perturbed percolation configuration ω_ϵ where only an ϵ-fraction of the edges has been resampled.

Remark The same statement holds for the more general crossing events described above (i.e., in $(n\Omega, n\partial_1, n\partial_2)$).

1.8 A dynamical consequence of noise sensitivity

One can consider a continuous time random walk $\{\omega_t\}_{t \geq 0}$ (implicitly depending on n which we suppress) on $\Omega_n := \{-1,1\}^n$ obtained by having each variable independently re-randomize at the times of a rate 1 Poisson process (so that the times between rerandomizations are independent exponential times with parameter 1). The stationary distribution is of course our usual probability measure, which is a product measure with 1 and -1 equally likely. Starting from stationarity, observe that the joint distribution of ω_s and ω_{s+t} is the same as the joint distribution of ω and ω_ϵ introduced earlier where $\epsilon = 1 - e^{-t}$.

Considering next a sequence of Boolean functions $\{f_n\}_{n\geq 1}$ mapping Ω_n into, say, $\{0,1\}$, we obtain a sequence of processes $\{g_n(t)\}$ defined by $g_n(t) := f_n(\omega_t)$. The following general result was proved in (BKS99) for the specific case of percolation crossings; however, their proof applies verbatim in this general context.

Theorem 1.23 (BKS99)   *Let $\{f_n\}_{n\geq 1}$ be a sequence of Boolean functions that is noise sensitive and satisfies $\delta_0 \leq \mathbb{P}(f_n(\omega) = 1) \leq 1 - \delta_0$ for all n for some $\delta_0 > 0$. Let S_n be the set of times in $[0,1]$ at which $g_n(t)$ changes its value. Then $|S_n| \to \infty$ in probability as $n \to \infty$.*

Proof   We first claim that for all $0 \leq a < b \leq 1$,

$$\lim_{n\to\infty} \mathbb{P}(S_n \cap [a,b] = \emptyset) = 0. \tag{1.3}$$

Let $W_{n,\epsilon} := \{\omega : \mathbb{P}[f_n(\omega_\epsilon) = 1 | \omega] \in [0, \delta_0/2] \cup [1 - \delta_0/2, 1]\}$. The noise sensitivity assumption, the (δ_0-)nondegenericity assumption, and Exercise 4.7 (which gives an alternative description of noise sensitivity) imply that for each $\epsilon > 0$,

$$\lim_{n\to\infty} \mathbb{P}(W_{n,\epsilon}) = 0.$$

Fix $\gamma > 0$ arbitrarily. Choose k so that $(1 - \delta_0/2)^k < \gamma/2$ and then choose $\epsilon := (b-a)/k$. Finally choose N so that for all $n \geq N$, $\mathbb{P}(W_{n,\epsilon}) \leq \gamma\delta_0/4$. Let $a = t_0 < t_1 < t_2 < \cdots < t_k = b$, where each $t_i - t_{i-1}$ equals $(b-a)/k$. We then have for $n \geq N$,

$$\mathbb{P}(S_n \cap [a,b] = \emptyset)$$
$$\leq \mathbb{P}(\omega_{t_{k-1}} \in W_{n,\epsilon}) + \mathbb{E}[I_{\{\omega_{t_{k-1}} \notin W_{n,\epsilon}\}} \mathbb{P}[S_n \cap [a,b] = \emptyset \mid \omega_{t_{k-1}}]]$$
$$\leq \frac{\gamma\delta_0}{4} \mathbb{E}[I_{\{\omega_{t_{k-1}} \notin W_{n,\epsilon}\}} \mathbb{P}[S_n \cap [t_{k-1},b] = \emptyset \mid \omega_{t_{k-1}}] \mathbb{P}[S_n \cap [a,t_{k-1}] = \emptyset \mid \omega_{t_{k-1}}]]$$

using the Markov property. If $\omega_{t_{k-1}} \notin W_{n,\epsilon}$, then

$$\mathbb{P}[S_n \cap [t_{k-1},b] = \emptyset \mid \omega_{t_{k-1}}] \leq \mathbb{P}[g_n(b) = g_n(t_{k-1}) \mid \omega_{t_{k-1}}] \leq 1 - \delta_0/2.$$

This yields

$$\mathbb{P}(S_n \cap [a,b] = \emptyset) \leq \frac{\gamma\delta_0}{4} + \left(1 - \frac{\delta_0}{2}\right)\mathbb{P}(S_n \cap [a,t_{k-1}] = \emptyset).$$

Continuing by induction $k - 1$ more times yields

$$\mathbb{P}(S_n \cap [a,b] = \emptyset) \leq \frac{\gamma}{2} + \left(1 - \frac{\delta_0}{2}\right)^k < \gamma.$$

As γ is arbitrary, this proves (1.3), which is the main step of the proof.

Now let $M \geq 1$ be an arbitrary integer and $\alpha > 0$. Partition $[0,1]$ into $2M$ intervals of length $\frac{1}{2M}$. By (1.3), we can choose N sufficiently large that for all $n \geq N$

$$\mathbb{P}(S_n \cap [0, \frac{1}{2M}] = \emptyset) \leq \alpha/2.$$

Let

$$X_n := \left| \{\ell \in \{1, \ldots, 2M\} : S_n \cap \left[\frac{\ell-1}{2M}, \frac{\ell}{2M} \right] = \emptyset \} \right|.$$

By stationarity and our choice of N, we have $\mathbb{E}[X_n] \leq \alpha M$ for all $n \geq N$. By Markov's inequality, we have for such n

$$\mathbb{P}(|S_n| \leq M) \leq \mathbb{P}(X_n \geq M) \leq \alpha.$$

As M and α are arbitrary, this implies that $|S_n| \to \infty$ in probability. $\square$

In Chapter 11, dynamical percolation is studied where one analyzes the dynamical model introduced in this section in the context of percolation on the full lattice.

1.9 Exercises

1.1 Determine the pivotal set (as a function of ω), the influence vector and the total influence for Examples 1.2–1.4.

1.2 Determine the influence vector for Iterated 3-Majority and Tribes.

1.3 Show that in Example 1.15 (Tribes) the variances stay bounded away from 0. If the blocks are taken to be of size $\log_2 n$ instead, show that the influences would all be of order $1/n$. Why does this not contradict the KKL theorem?

1.4 Ω_n has a graph structure where two elements are neighbors if they differ in exactly one location. The **edge boundary** of a subset $A \subseteq \Omega_n$, denoted by $\partial_E(A)$, is the set of edges where exactly one of the endpoints is in A.

Show that for any Boolean function, $\mathbf{I}(f) = |\partial_E(A_f)|/2^{n-1}$.

1.5 Prove Theorem 1.13. This is a type of Poincaré inequality. *Hint:* use the fact that $\mathrm{Var}(f)$ can be written $2\mathbb{P}[f(\omega) \neq f(\widetilde{\omega})]$, where $\omega, \widetilde{\omega}$ are independent and try to "interpolate" from ω to $\widetilde{\omega}$.

1.6 Show that Example 1.4 (Majority) is noise stable.

1.7   Prove that Example 1.5 (Iterated 3-Majority) is noise sensitive directly without relying on Theorem 1.21. *Hint:* Use the recursive structure of this example to show that the criterion of noise sensitivity is satisfied.

1.8   Prove that Example 1.15 (Tribes) is noise sensitive directly without using Theorem 1.21. Here there is no recursive structure, so a more "probabilistic" argument is needed.

1.9   Recall Example 1.6 (Clique containment).

(a) Prove that when $k_n = o(n^{1/2})$, $\mathbf{CLIQ}_n^{k_n}$ is asymptotically noise sensitive. *Hint:* Start by obtaining an upper bound on the influences (which are identical for each edge) using Exercise 1.4. Conclude by using Theorem 1.21.

(b) *(Challenging exercise).* Find a more direct proof of this fact (in the spirit of Exercise 1.8) which would avoid using Theorem 1.21.

As pointed out after Example 1.6, for most values of $k = k_n$, the Boolean function $\mathbf{CLIQ}_n^{k_n}$ becomes degenerate. The purpose of the rest of this exercise is to determine what the interesting regime is where $\mathbf{CLIQ}_n^{k_n}$ has a chance of being nondegenerate (i.e., the variances are bounded away from 0). The rest of this exercise is somewhat tangential to this chapter.

(c) If $1 \le k \le \binom{n}{2} = r$, what is the expected number of cliques in G_ω, $\omega \in \Omega_r$?

(d) Explain why there should be at most one choice of $k = k_n$ such that the variance of $\mathbf{CLIQ}_n^{k_n}$ remains bounded away from 0. (No rigorous proof is required.) Describe this choice of k_n. Check that it is indeed in the regime $2 \log_2(n)$.

(e) Note retrospectively that in fact, for any choice of $k = k_n$, $\mathbf{CLIQ}_n^{k_n}$ is noise sensitive.

1.10   Deduce from Theorem 1.21 that both Example 1.5 (Iterated 3-Majority) and Example 1.15 (Tribes) are noise sensitive.

1.11   Give a sequence of Boolean functions that is neither noise sensitive nor noise stable.

1.12   In the sense of Definition 1.17, show that for the Majority function and for fixed ϵ, any set of size $n^{1/2+\epsilon}$ has influence approaching 1 while any set of size $n^{1/2-\epsilon}$ has influence approaching 0.

1.13 Show that there exists $c > 0$ such that for any Boolean function

$$\sum_i \mathbf{I}_i^2(f) \geq c \operatorname{Var}^2(f)(\log^2 n)/n,$$

and show that this is sharp up to a constant. This result is also contained in (KKL88).

1.14 Do you think a "generic" Boolean function would be stable or sensitive? Justify your intuition. Show that if f_n is a "randomly" chosen Boolean function, then a.s. $\{f_n\}$ is noise sensitive. (It turns out that typical *monotone* Boolean functions are, by contrast, noise stable. This follows from the statement immediately after Theorem 1.2 in (Kor03), a large survey article concerning monotone Boolean functions.)

2

Percolation in a nutshell

To make this book as self-contained as possible, we review various aspects of the percolation model and give a short summary of the main useful results. For a complete account of percolation, see (Gri99), and for a more detailed study of the two-dimensional case, on which we concentrate here, see the lecture notes (Wer07).

2.1 The model

Let us start by introducing the model itself.

We are concerned mainly with two-dimensional percolation and we focus on two lattices: $\mathbb{Z}^2$ and the triangular lattice $\mathbb{T}$. However, all the results stated here for $\mathbb{Z}^2$ are also valid for percolation models on "reasonable" 2-d translation-invariant graphs for which the RSW theorem (see Section 2.2) is known to hold at the corresponding critical point.

Let us describe the model on the graph $\mathbb{Z}^2$, which has $\mathbb{Z}^2$ as its vertex set and edges between vertices having Euclidean distance 1. Let $\mathbb{E}^2$ denote the set of edges of the graph $\mathbb{Z}^2$. For any $p \in [0, 1]$ we define a random subgraph of $\mathbb{Z}^2$ as follows: independently for each edge $e \in \mathbb{E}^2$, we keep this edge with probability p and remove it with probability $1 - p$. Equivalently, this corresponds to defining a random configuration $\omega \in \{-1, 1\}^{\mathbb{E}^2}$ where, independently for each edge $e \in \mathbb{E}^2$, we declare the edge to be **open** ($\omega(e) = 1$) with probability p or **closed** ($\omega(e) = -1$) with probability $1 - p$. The law of the random subgraph (or configuration) defined in this way is denoted by $\mathbb{P}_p$.

Percolation is defined similarly on the triangular grid $\mathbb{T}$, except that on this lattice we instead consider *site* percolation (i.e., here we keep each site with probability p). The sites are the points $\mathbb{Z} + e^{i\pi/3}\mathbb{Z}$ so that neighboring sites have distance 1 from each other in the complex plane.

14

Figure 2.1 Pictures (by Oded Schramm) representing two
percolation configurations respectively on $\mathbb{T}$ and on $\mathbb{Z}^2$ (both at
$p = 1/2$). The sites of the triangular grid are represented by
hexagons.

2.2 Russo–Seymour–Welsh

We will often rely on the celebrated result known as the **RSW theorem**.

Theorem 2.1 (RSW; see (Gri99)) *For percolation on $\mathbb{Z}^2$ at $p = 1/2$, one
has the following property concerning crossing events. Let $a, b > 0$. There
exists a constant $c = c(a,b) > 0$, such that for any $n \geq 1$, if A_n denotes
the event that there is a left-to-right crossing in the rectangle $([0, a \cdot n]
\times [0, b \cdot n]) \cap \mathbb{Z}^2$, then*

$$c < \mathbb{P}_{1/2}[A_n] < 1 - c.$$

*In other words, this says that the Boolean functions f_n defined in Example
1.22 of Chapter 1 are nondegenerate.*

The same result holds in the case of site percolation on $\mathbb{T}$ (also at $p =
1/2$).

The parameter $p = 1/2$ plays a very special role in the two models under
consideration. Indeed, there is a natural way to associate to each percola-
tion configuration $\omega_p \sim \mathbb{P}_p$ a **dual configuration** ω_{p^*} on the so-called **dual
graph**. In the case of $\mathbb{Z}^2$, its dual graph $(\mathbb{Z}^2)^*$ can be realized as $\mathbb{Z}^2 + (\frac{1}{2}, \frac{1}{2})$.

See Figure 2.2 for an illustration of this. In the case of the triangular lattice, $\mathbb{T}^* = \mathbb{T}$. It is easy to see that in both cases $p^* = 1 - p$. Hence, at $p = 1/2$, our two models happen to be *self-dual*.

Figure 2.2 The lattice $\mathbb{Z}^2$ and its (self-)dual graph.

This duality has the following very important consequence. For a domain in $\mathbb{T}$ with two specified boundary arcs, there is a left–right crossing of white hexagons if and only if there is no top–bottom crossing of black hexagons.

2.3 Phase transition

Percolation theory concerns large-scale connectivity properties of the random configuration $\omega = \omega_p$. In particular, as the level p rises above a certain critical parameter $p_c(\mathbb{Z}^2)$, an infinite cluster (almost surely) emerges. This is the well-known *phase transition* of percolation. (We will often use the expression "there is percolation" to mean that there is an infinite connected component.) By a famous theorem of Kesten, it is the case that $p_c(\mathbb{Z}^2) = 1/2$. On the triangular grid, one also has $p_c(\mathbb{T}) = 1/2$.

Let $\{0 \overset{\omega}{\longleftrightarrow} \infty\}$ denote the event that there exists a self-avoiding path from 0 to ∞ consisting of open edges. This phase transition can be measured with the *density function* $\theta_{\mathbb{Z}^2}(p) := \mathbb{P}_p(0 \overset{\omega}{\longleftrightarrow} \infty)$ which encodes important properties of the large-scale connectivities of the random configuration ω: it corresponds to the density averaged over the space $\mathbb{Z}^2$ of the (almost surely unique) infinite cluster. The shape of the function $\theta_{\mathbb{Z}^2}$ is pictured in Figure 2.3. Note in particular the infinite derivative at p_c.

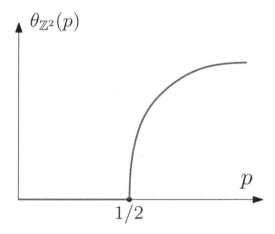

Figure 2.3 The density function $p \mapsto \theta_{\mathbb{Z}^2}(p)$.

2.4 Conformal invariance at criticality and the Schramm–Loewner evolution (SLE)

It has been conjectured for a long time that percolation should be *asymptotically* conformally invariant at the critical point. This should be understood in the same way as the fact that a Brownian motion (ignoring its time-parametrization) is a conformally invariant probabilistic object. Letting $\mathbb{D}$ denote the unit disk, one way to picture this conformal invariance is as follows: consider the "largest" cluster C_δ surrounding 0 in $\delta\mathbb{Z}^2 \cap \mathbb{D}$ and such that $C_\delta \cap \partial\mathbb{D} = \emptyset$. Now consider some other simply connected domain Ω containing 0. Let $\hat{C}_\delta$ be the largest cluster surrounding 0 in a critical configuration in $\delta\mathbb{Z}^2 \cap \Omega$ and such that $\hat{C}_\delta \cap \partial\Omega = \emptyset$. Now let ϕ be the conformal map from $\mathbb{D}$ to Ω such that $\phi(0) = 0$ and $\phi'(0) > 0$. Even though the random sets $\phi(C_\delta)$ and $\hat{C}_\delta$ do not lie on the same lattice, the conformal invariance principle claims that when $\delta = o(1)$, these two random clusters are very close in law.

Over the last fifteen years, two major breakthroughs have enabled researchers to obtain a much better understanding of the critical regime of percolation:

- The invention of the SLE processes by Oded Schramm (Sch00),
- The proof of conformal invariance on $\mathbb{T}$ by Stanislav Smirnov (Smi01).

The simplest precise statement concerning conformal invariance is the following. Let Ω be a bounded simply connected domain of the plane and let A, B, C, and D be four points on the boundary of Ω in clockwise order. Scale the hexagonal lattice $\mathbb{T}$ by $1/n$ and perform critical percolation on this scaled lattice. Let $\mathbb{P}(\Omega, A, B, C, D, n)$ denote the probability that in the $1/n$ scaled hexagonal lattice there is an open path of hexagons in Ω going from the boundary of Ω between A and B to the boundary of Ω between C and D.

Theorem 2.2 *(Smirnov, (Smi01))*

(i). For all Ω and A, B, C, and D as above,

$$\mathbb{P}(\Omega, A, B, C, D, \infty) := \lim_{n \to \infty} \mathbb{P}(\Omega, A, B, C, D, n)$$

exists and is conformally invariant in the sense that if f is a conformal mapping, then $\mathbb{P}(\Omega, A, B, C, D, \infty) = \mathbb{P}(f(\Omega), f(A), f(B), f(C), f(D), \infty)$.

(ii). If Ω is an equilateral triangle (with side lengths 1), A, B, and C the three corner points and D on the line between C and A having distance x from C, then the above limiting probability is x. (Observe, by conformal invariance, that this gives the limiting probability for all domains and four boundary points.)

Part (i) of the theorem was conjectured by M. Aizenman, while J. Cardy conjectured the limit for the case where Ω is a rectangle using the four corners. In this case, the formula is quite complicated involving hypergeometric functions but Lennart Carleson then realized that, assuming conformal invariance, this is equivalent to the simpler formula given in part (ii) for a triangle.

On $\mathbb{Z}^2$ at $p_c = 1/2$, proving conformal invariance is still a challenging open problem.

We do not define the SLE processes in this book. There are many sources, for example (B12) and the references contained therein. The illustration below explains how SLE curves arise naturally in the percolation picture.

This celebrated picture (by Oded Schramm) represents an **exploration path** on the triangular lattice. This exploration path, which turns right when encountering gray hexagons and left when encountering white ones, asymptotically converges toward SLE$_6$ (as the mesh size goes to 0).

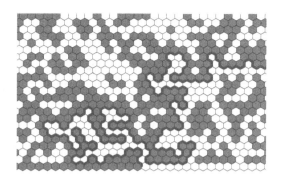

2.5 Critical exponents

The proof of conformal invariance combined with the detailed information given by the SLE$_6$ process provides very precise information about the critical and *near-critical* behavior of site percolation on $\mathbb{T}$. For instance, it can be shown that on the triangular lattice the density function $\theta_{\mathbb{T}}(p)$ exhibits the following behavior near $p_c = 1/2$:

$$\theta(p) = (p - 1/2)^{\frac{5}{36} + o(1)}$$

when $p \to 1/2+$ (see (Wer07)).

In the chapters that follow, we often rely on three types of percolation events: *one-arm two-arm*, and *four-arm* events. They are defined as follows: for any radius $R > 1$, let A_R^1 be the event that the site 0 is connected to distance R by some open path (one "arm"). Next, let A_R^2 be the event that there are two arms of different colors from the site 0 (which itself can be of either color) to distance R away. Finally, let A_R^4 be the event that there are four arms of alternating color from the site 0 (which itself can be of either color) to distance R away (i.e., there are four connected paths, two open, two closed from 0 to radius R and they occur in alternating order). See Figure 2.4 for a realization of two of these events.

It was proved in (LSW02) that the probability of the one-arm event decays as follows:

$$\mathbb{P}[A_R^1] := \alpha_1(R) = R^{-\frac{5}{48} + o(1)}.$$

For the two-arm and four-arm events (as well as many other similarly defined events), it was proved by Smirnov and Werner in (SW01) that these

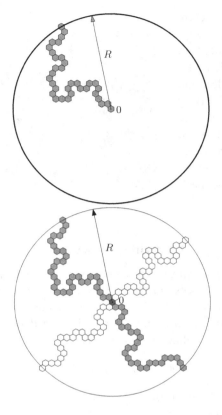

Figure 2.4 A realization of the *one-arm* event is pictured on the top; the *four-arm* event is pictured on the bottom.

probabilities decay as follows:

$$\mathbb{P}[A_R^2] := \alpha_2(R) = R^{-\frac{1}{4}+o(1)}$$

and

$$\mathbb{P}[A_R^4] := \alpha_4(R) = R^{-\frac{5}{4}+o(1)}.$$

The four exponents we encounter concerning $\theta_{\mathbb{T}}$, α_1, α_2, and α_4 (i.e, $\frac{5}{36}$, $\frac{5}{48}$, $\frac{1}{4}$, and $\frac{5}{4}$) are known as *critical exponents*. The *four-arm* event is clearly of particular relevance to us here. Indeed, if a point x is in the "bulk" of a domain $(n\Omega, n\partial_1, n\partial_2)$, the event that this point is pivotal for the left–right crossing event A_n is intimately related to the four-arm event. See Chapter 6 for more details.

The presence of $o(1)$ terms in the preceding statements (which of course go to zero as $R \to \infty$) reveals that the critical exponents are known so far only up to "logarithmic" corrections. It is conjectured that there are no such "logarithmic" corrections, but at the moment one has to deal with their possible existence. More specifically, it is believed that for the one-arm event,

$$\alpha_1(R) \asymp R^{-\frac{5}{48}},$$

where $\asymp$ means that the ratio of the two sides is bounded away from 0 and ∞ uniformly in R; similarly for the other arm events.

2.6 Quasi-multiplicativity

Finally, let us end this overview with a type of scale-invariance property of these arm events. More precisely, it is often convenient to "divide" these arm events into different scales. For this purpose, we introduce $\alpha_4(r,R)$ (with $r \le R$) as the probability that the four-arm event holds from radius r to radius R ($\alpha_1(r,R)$, $\alpha_2(r,R)$ and $\alpha_3(r,R)$ are defined analogously). By independence on disjoint sets, it is clear that if $r_1 \le r_2 \le r_3$ then we have $\alpha_4(r_1,r_3) \le \alpha_4(r_1,r_2)\alpha_4(r_2,r_3)$. A very useful property known as **quasi-multiplicativity** claims that up to constants, these two expressions are the same (this makes the division into several scales practical). This property can be stated as follows.

Proposition 2.3 (quasi-multiplicativity, (Kes87)) *For any $r_1 \le r_2 \le r_3$, one has, for both $\mathbb{Z}^2$ and $\mathbb{T}$,*

$$\alpha_4(r_1,r_3) \asymp \alpha_4(r_1,r_2)\alpha_4(r_2,r_3).$$

See (Wer07; Nol09; SS10) for more details. The same property holds for the one-arm event but this is much easier to prove: it is an easy consequence of the RSW Theorem 2.1 and the so-called Fortuin–Kasteleyn–Ginibre (FKG) inequality, which says that increasing events are positively correlated. Readers might consider doing this as an exercise.

3

Sharp thresholds and the critical point for 2-d percolation

In this chapter, we concentrate on monotone Boolean functions where the total influence corresponds to the derivative of the probability of the event in question with respect to the parameter (this is the Margulis–Russo formula). This, together with general versions of Theorems 1.14 and 1.16 for general parameters p, will allow us to prove a sharp threshold result due to Friedgut and Kalai as well as to prove Kesten's theorem that the critical value for percolation is $1/2$.

3.1 Monotone functions and the Margulis–Russo formula

The class of so-called monotone functions plays a central role in this subject.

Definition 3.1 A function f is **monotone** if $x \leq y$ (meaning $x_i \leq y_i$ for each i) implies that $f(x) \leq f(y)$. An event is monotone if its indicator function is monotone.

Recall that when the underlying variables are independent with the value 1 having probability p, we let $\mathbb{P}_p$ and $\mathbb{E}_p$ denote probabilities and expectations.

It is fairly obvious that for f monotone, $\mathbb{E}_p(f)$ should be increasing in p. The Margulis–Russo formula gives us an explicit formula for this (nonnegative) derivative.

Theorem 3.2 *Let A be an increasing event in* Ω_n. *Then*

$$d(\mathbb{P}_p(A))/dp = \sum_i \mathbf{I}_i^p(A).$$

Proof Allow each variable x_i to have its own parameter p_i and let $\mathbb{P}_{p_1,\dots,p_n}$ and $\mathbb{E}_{p_1,\dots,p_n}$ be the corresponding probability measure and expectation. It suffices to show that

$$\partial(\mathbb{P}_{p_1,\dots,p_n}(A))/\partial p_i = \mathbf{I}_i^{p_1,\dots,p_n}(A),$$

where the definition of this latter term is clear. Without loss of generality, take $i = 1$. Now

$$\mathbb{P}_{p_1,\dots,p_n}(A) = \mathbb{P}_{p_1,\dots,p_n}(A \backslash \{1 \in \mathcal{P}_A\}) + \mathbb{P}_{p_1,\dots,p_n}(A \cap \{1 \in \mathcal{P}_A\}).$$

The event in the first term on the right-hand side is measurable with respect to the other variables and hence the first term does not depend on p_1 while the second term is

$$p_1 \mathbb{P}_{p_2,\dots,p_n}(\{1 \in \mathcal{P}_A\}),$$

since $A \cap \{1 \in \mathcal{P}_A\}$ is the event $\{x_1 = 1\} \cap \{1 \in \mathcal{P}_A\}$. $\qquad\square$

3.2 KKL away from the uniform measure case

Recall now Theorem 1.14. For sharp threshold results, one needs lower bounds on the total influence not just at the special parameter $1/2$ but at all p.

We want at our disposal two main results concerning the KKL result for general p. The proofs of these theorems are outlined in the exercises in Chapter 5. We also refer to (Ros08).

Theorem 3.3 (BKK$^+$92) *There exists a universal $c > 0$ such that for any Boolean function f mapping Ω_n into $\{0, 1\}$ and for any p, there exists some i such that*

$$\mathbf{I}_i^p(f) \geq c \mathrm{Var}_p(f)(\log n)/n.$$

Theorem 3.4 (BKK$^+$92) *There exists a universal $c > 0$ such that for any Boolean function f mapping Ω_n into $\{0, 1\}$ and for any p,*

$$\mathbf{I}^p(f) \geq c \mathrm{Var}_p(f) \log(1/\delta_p),$$

where $\delta_p := \max_i \mathbf{I}_i^p(f)$.

3.3 Sharp thresholds in general: The Friedgut–Kalai theorem

Theorem 3.5 (FK96) *There exists a constant $c_1 < \infty$ such that for any monotone event A on n variables where all the influences are the same, if $\mathbb{P}_{p_1}(A) > \epsilon$, then*

$$\mathbb{P}_{p_1 + \frac{c_1 \log(1/(2\epsilon))}{\log n}}(A) > 1 - \epsilon.$$

Remark This theorem says that for fixed ϵ, the probability of A moves from below ϵ to above $1 - \epsilon$ in an interval of p of length of order at most $1/\log n$. The assumption of equal influences holds, for example, if the event is invariant under some transitive action, which is often the case. For example, it holds for Example 1.5 (Iterated 3-Majority) as well as for any (monotone) graph property, such as connectivity, in the context of the so-called Erdős–Rényi random graph model $\mathcal{G}(n,p)$, where a random graph on n vertices is chosen by independently putting in each potential edge with probability p.

Proof of Theorem 3.5. Theorem 3.3 and all the influences being the same tell us that

$$I^p(A) \geq c \min\{\mathbb{P}_p(A), 1 - \mathbb{P}_p(A)\} \log n$$

for some $c > 0$. Hence Theorem 3.2 yields

$$d(\log(\mathbb{P}_p(A)))/dp \geq c \log n$$

if $\mathbb{P}_p(A) \leq 1/2$. Letting $p^* := p_1 + \frac{\log(1/2\epsilon)}{c\log n}$, an easy computation (using the fundamental theorem of calculus) yields

$$\log(P_{p^*}(A)) \geq \log(1/2).$$

Next, if $\mathbb{P}_p(A) \geq 1/2$, then

$$d(\log(1 - \mathbb{P}_p(A)))/dp \leq -c \log n$$

from which another application of the fundamental theorem yields

$$\log(1 - P_{p^{**}}(A)) \leq -\log(1/\epsilon)$$

where $p^{**} := p^* + \frac{\log(1/2\epsilon)}{c\log n}$. Letting $c_1 = 2/c$ gives the result. $\qquad\square$

3.4 The critical point for percolation for $\mathbb{Z}^2$ and $\mathbb{T}$ is $\frac{1}{2}$

Theorem 3.6 (Kes80)

$$p_c(\mathbb{Z}^2) = p_c(\mathbb{T}) = 1/2.$$

Proof We first show that $\theta(1/2) = 0$. Let $\text{Ann}(\ell) := [-3\ell, 3\ell]\backslash[-\ell, \ell]$ and C_k be the event that there is a circuit in $\text{Ann}(4^k) + 1/2$ in the dual lattice around the origin consisting of closed edges. The C_k are independent and RSW and FKG show that for some $c > 0$, $\mathbb{P}_{1/2}(C_k) \geq c$ for all k. This gives that $\mathbb{P}_{1/2}(C_k$ infinitely often$) = 1$ and hence $\theta(1/2) = 0$.

The next key step is a *finite size criterion*, which implies percolation and which is interesting in itself. We outline its proof afterwards.

Proposition 3.7   *(Finite size criterion) Let J_n be the event that there is a crossing of a $2n \times (n-2)$ box. For any p, if there exists n such that*

$$\mathbb{P}_p(J_n) \geq 0.98,$$

then a.s. there exists an infinite cluster at parameter p.

    Assume now that $p_c = 1/2 + \delta$ with $\delta > 0$. Let $I = [1/2, 1/2 + \delta/2]$. Since $\theta(1/2 + \delta/2) = 0$, it is easy to see that the maximum influence over all variables and over all $p \in I$ goes to 0 with n because being pivotal implies the existence of an open path from a neighbor of the given edge to distance $n/2$ away. Next, by RSW, $\inf_n \mathbb{P}_{1/2}(J_n) > 0$. If for all n, $\mathbb{P}_{1/2+\delta/2}(J_n) < 0.98$, then Theorems 3.2 and 3.4 would allow us to conclude that the derivative of $\mathbb{P}_p(J_n)$ goes to ∞ uniformly on I as $n \to \infty$, which is a contradiction. Hence $\mathbb{P}_{1/2+\delta/2}(J_n) \geq 0.98$ for some n, implying, by Proposition 3.7, that $\theta(1/2 + \delta/2) > 0$, a contradiction.   □

Outline of proof of Proposition 3.7. The first step is to show that for any p and for any $\epsilon \leq 0.02$, if $\mathbb{P}_p(J_n) \geq 1 - \epsilon$, then $\mathbb{P}_p(J_{2n}) \geq 1 - \epsilon/2$. To see this, we first, by FKG and "gluing," observe that we can cross a $4n \times (n-2)$ box with probability at least $1 - 5\epsilon$ and hence one obtains that $\mathbb{P}_p(J_{2n}) \geq 1 - \epsilon/2$ because, for this event to fail, it must fail in both the top and bottom halves of the box. It then follows that if we place a sequence of (possibly rotated and translated) boxes of sizes $2^{n+1} \times 2^n$ anywhere, then, with probability 1, all but finitely many are crossed. Finally, we can place these boxes in an intelligent way such that crossing all but finitely many of them necessarily entails the existence of an infinite cluster (see Figure 3.1).   □

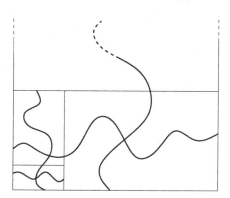

Figure 3.1 Creating an infinite cluster using box crossings.

3.5 Further discussion

The Margulis–Russo formula is due independently to Margulis (Mar74) and Russo (Rus81).

Bollobás and Riordan had the idea to use results from (KKL88) to show that $p_c = 1/2$ (see (BR06a)); it was understood much earlier that obtaining a sharp threshold was the key step in proving that $p_c(\mathbb{Z}^2) = 1/2$. Kesten (Kes80) showed the necessary sharp threshold by obtaining a lower bound on the expected number of pivotals in a hands-on fashion. In the proof given here, we used Theorem 3.4. However, in (BR06a), Theorem 3.5, which uses only Theorem 3.3 and not Theorem 3.4, was used. To use Theorem 3.5, it was necessary to symmetrize things by placing the model on a torus instead. Russo (Rus82) had developed an earlier qualitative "threshold" result and demonstrated how this also sufficed to show that $p_c = 1/2$. We state this result.

Theorem 3.8 (Rus82)   *For all $\epsilon > 0$, there exists $\delta > 0$ such that if A is any increasing event satisfying $\mathbf{I}_i^p(A) < \delta$ for all p and all i, then there exists p_0 such that (i) $\mathbb{P}_p(A) \leq \epsilon$ for all $p \leq p_0 - \epsilon$ and (ii) $\mathbb{P}_p(A) \geq 1 - \epsilon$ for all $p \geq p_0 + \epsilon$.*

Remarks

1. Observe that Theorem 3.8 follows immediately from Theorems 3.2 and 3.4.
2. However, this result of course predates the theorems in (KKL88). Moreover, the proof is based purely on (nontrivial) hands-on probabilistic arguments unlike those in (KKL88).
3. We could have appealed to this theorem instead to derive the critical value for percolation.
4. We mention finally that Theorem 5.4 (to follow) yields a stronger and more quantitative version of Theorem 3.8 but its proof uses more sophisticated methods.

3.6 Exercises

3.1   Develop an alternative proof of the Margulis–Russo formula using classical couplings.

3.2   Study, as best as you can, the "threshold windows" in the following examples (i.e., where and how long does it take to go from a probability of order ϵ to a probability of order $1 - \epsilon$):

(a) for **DICT**$_n$,
(b) for **MAJ**$_n$,
(c) for the Tribes example,
(d) for the Iterated 3-Majority example.

Do not rely on a (KKL88) type of result, but instead do hands-on computations specific to each case.

3.3 Write out the details of the proof of Proposition 3.7 as outlined in the chapter.

3.4 Prove Proposition 3.7 using a different approach based on a renormalization argument together with the **Peierls argument** (this is the approach used, for example, in (BR06b)).

First renormalize the lattice $\mathbb{Z}^2$ by using copies of $3n \times n$ rectangles as in the figure. On the renormalized lattice, declare each edge $e = (u, v)$ to be open if the corresponding rectangle is crossed along the direction of e. One obtains a **correlated** percolation model on $\mathbb{Z}^2$ with parameter $\bar{p} = \mathbb{P}_p[J_n] \geq 0.98$.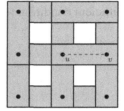

The goal here is to show that despite the dependence between edges in this percolation model, there is almost surely an infinite cluster in the renormalized percolation configuration.

(a) Show that the percolation model thus constructed is a model of **1-dependent** percolation in the sense that if F and G are any two sets of edges in $\mathbb{Z}^2$, then the states of edges in F are independent of the states of edges in G if and only if none of the edges in F have a common vertex with an edge in G. (More generally a k-independent percolation model has the same independence property for sets of edges satisfying $d_{\text{graph}}(F, G) \geq k$.)
(b) Prove that there is an infinite cluster with positive probability by following the classical proof that $\theta(p) > 0$ in $\mathbb{Z}^2$ when p is sufficiently close to 1; i.e., use a duality argument. The difficulty here is handling the 1-dependence between edges (and dual edges). To overcome this difficulty, note that if γ_{2n} is a self-avoiding path of length $2n$ in the dual lattice, then by removing every other edge, one obtains a simple enough combinatorial object made of independent edges. (See (BR06b) for a detailed proof.)

3.5 (*What is the "sharpest" monotone event?*)   Show that among all monotone Boolean functions on Ω_n, $\mathbf{MAJ}_n$ is the function with largest total influence (at $p = 1/2$).

Hint: Use the Margulis–Russo formula.

3.6   A consequence of Exercise 3.5 is that the total influence at $p = 1/2$ of any monotone function is at most $O(\sqrt{n})$. A similar argument shows that for any p, there is a constant C_p so that the total influence at level p of any monotone function is at most $C_p \sqrt{n}$. Prove nonetheless that there exists $c > 0$ such that for any n, there exists a monotone function $f = f_n$ and $p = p_n$ so that the total influence of f at level p is at least cn.

3.7   Find a monotone function $f \colon \Omega_n \to \{0, 1\}$ such that $d(\mathbb{E}_p(f))/dp$ is very large at $p = 1/2$, but nevertheless there is no sharp threshold for f (this means that a large total influence at some value of p is not in general a sufficient condition for a sharp threshold).

4

Fourier analysis of Boolean functions
(first facts)

In this chapter, we introduce the key technical tool for our study, which is
Fourier analysis on the hypercube; this allows us to define the spectrum of
a Boolean function. From here, we are able to give a spectral characterization of both noise sensitivity as well as noise stability. We also see two
important relationships between the notion of influence and the spectrum.

4.1 Discrete Fourier analysis and the energy spectrum

To understand and analyze the concepts introduced thus far, which are in
some sense purely probabilistic, a critical tool is Fourier analysis on the
hypercube.

Recall that our Boolean functions map the hypercube $\Omega_n := \{-1, 1\}^n$ into
$\{-1, 1\}$ or $\{0, 1\}$, where Ω_n is endowed with the uniform measure $\mathbb{P} = \mathbb{P}^n = (\frac{1}{2}\delta_{-1} + \frac{1}{2}\delta_1)^{\otimes n}$.

To apply Fourier analysis, the natural setup is to enlarge our discrete
space of Boolean functions and to consider instead the larger space
$L^2(\{-1, 1\}^n)$ of real-valued functions on Ω_n endowed with the inner product

$$\langle f, g \rangle := \sum_{x_1, \ldots, x_n} 2^{-n} f(x_1, \ldots, x_n) g(x_1, \ldots, x_n)$$

$$= \mathbb{E}[fg] \text{ for all } f, g \in L^2(\Omega_n),$$

where $\mathbb{E}$ denotes expectation with respect to the uniform measure $\mathbb{P}$ on Ω_n.

For any subset $S \subseteq \{1, 2 \ldots, n\}$, let χ_S be the function on $\{-1, 1\}^n$ defined
for any $x = (x_1, \ldots, x_n)$ by

$$\chi_S(x) := \prod_{i \in S} x_i. \tag{4.1}$$

Note that $\chi_\emptyset \equiv 1$. It is straightforward (check this!) to see that this family of
2^n functions forms an orthonormal basis for $L^2(\{-1, 1\}^n)$. Thus, any function f on Ω_n (and a fortiori any Boolean function f) can be decomposed

as

$$f = \sum_{S \subseteq \{1,\ldots,n\}} \hat{f}(S) \chi_S,$$

where $\{\hat{f}(S)\}_{S \subseteq [n]}$ are the so-called Fourier coefficients of f. They are also sometimes called the **Fourier–Walsh** coefficients of f and they satisfy

$$\hat{f}(S) := \langle f, \chi_S \rangle = \mathbb{E}[f \chi_S].$$

Note that $\hat{f}(\emptyset)$ is the average $\mathbb{E}[f]$ and we have Parseval's formula, which states that

$$\mathbb{E}(f^2) = \sum_{S \subseteq \{1,\ldots,n\}} \hat{f}^2(S).$$

As in classical Fourier analysis, if f is some Boolean function, its Fourier–Walsh coefficients provide information on the "regularity" of f. We sometimes use the term *spectrum* when referring to the set of Fourier coefficients.

Of course many other orthonormal bases exist for $L^2(\{-1,1\}^n)$, but the particular set of functions $\{\chi_S\}_{S \subseteq \{1,\ldots,n\}}$ arises naturally in many situations. First of all there is a well-known theory of Fourier analysis on groups, a theory that is particularly simple and elegant on abelian groups (thus including our special case of $\{-1,1\}^n$, but also $\mathbb{R}/\mathbb{Z}$, $\mathbb{R}$ and so on). For abelian groups, what turns out to be relevant for harmonic analysis is the set $\hat{G}$ of **characters** of G (i.e., the set of group homomorphisms from G to $\mathbb{C}^*$). In our case of $G = \{-1,1\}^n$, the characters are precisely our functions χ_S indexed by $S \subseteq \{1,\ldots,n\}$, because they satisfy $\chi_S(x \cdot y) = \chi_S(x) \chi_S(y)$. This background is not needed, however, and we won't speak in these terms.

The functions $\{\chi_S\}_{S \subseteq \{1,\ldots,n\}}$ also arise naturally if one performs a simple random walk on the hypercube (equipped with the Hamming graph structure), since they are the eigenfunctions of the corresponding Markov chain (heat kernel) on $\{-1,1\}^n$. This is why we will see later in this chapter that the basis $\{\chi_S\}$ is particularly well adapted to our study of noise sensitivity.

We introduce one more concept here without motivation; it will be very well motivated later in the chapter.

Definition 4.1   For any real-valued function $f : \Omega_n \to \mathbb{R}$, the **energy spectrum** E_f is defined by

$$E_f(m) := \sum_{|S|=m} \hat{f}(S)^2, \quad m \in \{1,\ldots,n\}.$$

4.2 Examples

First note that, from the Fourier point of view, the Dictator and Parity functions have simple representations because they are χ_1 and $\chi_{[n]}$, respectively. Each of the two corresponding energy spectra are trivially concentrated on one point, namely 1 and n.

For Example 1.4 (Majority), Bernasconi explicitly computed the Fourier coefficients and, when n goes to infinity, one obtains the following asymptotic formula for the energy spectrum:

$$E_{\mathbf{MAJ}_n}(m) = \sum_{|S|=m} \widehat{\mathbf{MAJ}}_n(S)^2 = \begin{cases} \frac{4}{\pi m 2^m} \binom{m-1}{\frac{m-1}{2}} + O(m/n) & \text{if } m \text{ is odd}, \\ 0 & \text{if } m \text{ is even}. \end{cases}$$

(Readers might think about why the "even" coefficients are 0.) See (O'D03b) for a nice overview and references concerning the spectral behavior of the Majority function.

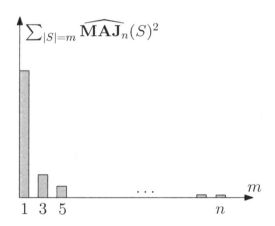

Figure 4.1 Shape of the energy spectrum for the Majority function.

Figure 4.1 represents the shape of the energy spectrum of $\mathbf{MAJ}_n$, which is concentrated on low frequencies, as is typical of stable functions.

4.3 Noise sensitivity and stability in terms of the energy spectrum

We now describe the concepts of noise sensitivity and noise stability in terms of the energy spectrum.

The first step is to note that, given any real-valued function $f : \Omega_n \to \mathbb{R}$, the correlation between $f(\omega)$ and $f(\omega_\epsilon)$ is nicely expressed in terms of the Fourier coefficients of f as follows:

$$
\begin{aligned}
\mathbb{E}[f(\omega)f(\omega_\epsilon)] &= \mathbb{E}\Big[\Big(\sum_{S_1} \hat{f}(S_1)\chi_{S_1}(\omega)\Big)\Big(\sum_{S_2}\hat{f}(S_2)\chi_{S_2}(\omega_\epsilon)\Big)\Big] \\
&= \sum_S \hat{f}(S)^2 \mathbb{E}[\chi_S(\omega)\chi_S(\omega_\epsilon)] \\
&= \sum_S \hat{f}(S)^2(1-\epsilon)^{|S|}.
\end{aligned}
\tag{4.2}
$$

Moreover, we immediately obtain

$$
\mathrm{Cov}(f(\omega), f(\omega_\epsilon)) = \sum_{m=1}^{n} E_f(m)(1-\epsilon)^m.
\tag{4.3}
$$

Note that either of (4.2) and (4.3) tells us that $\mathrm{Cov}(f(\omega), f(\omega_\epsilon))$ is nonnegative and decreasing in the randomization probability ϵ. Also, we see that the "level of noise sensitivity" of a Boolean function is naturally encoded in its energy spectrum. It is now an easy exercise to prove the following proposition.

Proposition 4.2 (BKS99)   *A sequence of Boolean functions* $f_n \colon \{-1, 1\}^{m_n} \to \{0, 1\}$ *is noise sensitive if and only if, for any* $k \geq 1$,

$$
\sum_{m=1}^{k}\sum_{|S|=m} \hat{f}_n(S)^2 = \sum_{m=1}^{k} E_{f_n}(m) \xrightarrow[n\to\infty]{} 0.
$$

Moreover, (1.1) holding does not depend on the value of $\epsilon \in (0, 1)$ *chosen.*

Deriving the similar spectral description of noise stability given (4.2) is an easy exercise.

Proposition 4.3 (BKS99)   *A sequence of Boolean functions* $f_n \colon \{-1, 1\}^{m_n} \to \{0, 1\}$ *is noise stable if and only if, for any* $\epsilon > 0$, *there exists k such that for all n,*

$$
\sum_{m=k}^{\infty}\sum_{|S|=m} \hat{f}_n(S)^2 = \sum_{m=k}^{\infty} E_{f_n}(m) < \epsilon.
$$

So, as argued in the Preface, a function of "high frequency" will be sensitive to noise while a function of "low frequency" will be stable.

4.4 Link between the spectrum and influence

In this section, we relate the notion of influence to that of the spectrum.

Proposition 4.4 *If $f: \Omega_n \to \{0,1\}$, then for all k,*

$$\mathbf{I}_k(f) = 4 \sum_{S:k\in S} \hat{f}(S)^2$$

and

$$\mathbf{I}(f) = 4 \sum_{S} |S| \hat{f}(S)^2.$$

Proof For $f: \Omega_n \to \mathbb{R}$, we introduce the functions

$$\nabla_k f: \begin{cases} \Omega_n & \to & \mathbb{R} \\ \omega & \mapsto & f(\omega) - f(\sigma_k(\omega)) \end{cases} \quad \text{for all } k \in [n],$$

where σ_k maps Ω_n to itself by flipping the kth bit (thus $\nabla_k f$ corresponds to a discrete derivative along the kth bit).

Observe that

$$\nabla_k f(\omega) = \sum_{S\subseteq\{1,\dots,n\}} \hat{f}(S)[\chi_S(\omega) - \chi_S(\sigma_k(\omega))] = \sum_{S\subseteq\{1,\dots,n\},k\in S} 2\hat{f}(S)\chi_S(\omega),$$

from which it follows that for any $S \subseteq [n]$,

$$\widehat{\nabla_k f}(S) = \begin{cases} 2\hat{f}(S) & \text{if } k\in S, \\ 0 & \text{otherwise.} \end{cases} \tag{4.4}$$

Clearly, if f maps into $\{0,1\}$, then $\mathbf{I}_k(f) := \|\nabla_k f\|_1$ and since $\nabla_k f$ takes values in $\{-1,0,1\}$ in this case, we have $\|\nabla_k f\|_1 = \|\nabla_k f\|_2^2$. Applying Parseval's formula to $\nabla_k f$ and using (4.4), we obtain the first statement of the proposition. The second is obtained by summing over k and exchanging the order of summation. □

Remark If f maps into $\{-1,1\}$ instead, then one can easily check that $\mathbf{I}_k(f) = \sum_{S:k\in S} \hat{f}(S)^2$ and $\mathbf{I}(f) = \sum_S |S| \hat{f}(S)^2$.

4.5 Monotone functions and their spectra

Monotone functions enjoy an alternative useful spectral description for their influences.

Proposition 4.5 *If $f: \Omega_n \to \{0,1\}$ is monotone, then for all k*

$$\mathbf{I}_k(f) = 2\hat{f}(\{k\}).$$

If f maps into $\{-1, 1\}$ instead, then $\mathbf{I}_k(f) = \hat{f}(\{k\})$. (The Parity example in Chapter 1 shows that the assumption of monotonicity is needed here; the proof also shows that the weaker result with = replaced by $\geq$ holds in general.)

Proof We prove only the first statement; the second is proved in the same way.

$$\hat{f}(\{k\}) := \mathbb{E}[f\chi_{\{k\}}] = \mathbb{E}[f\chi_{\{k\}}I_{\{k \notin \mathcal{P}\}}] + \mathbb{E}[f\chi_{\{k\}}I_{\{k \in \mathcal{P}\}}].$$

It is easily seen that the first term on the right-hand side is 0 (whether f is monotone or not) and the second term is $\frac{\mathbf{I}_k(f)}{2}$ due to monotonicity.   □

Remarks Proposition 4.5 tells us that, for monotone functions mapping into $\{-1, 1\}$, the sum in Theorem 1.21 is exactly the total weight of the level-1 Fourier coefficients, that is, the energy spectrum at 1, $E_f(1)$. (If we map into $\{0, 1\}$ instead, there is simply an extra irrelevant factor of 4.) So Theorem 1.21 and Propositions 4.2 and 4.5 imply that for monotone functions, if the energy spectrum at 1 goes to 0, then this is true for any fixed level. In addition, Propositions 4.2 (with $k = 1$) and 4.5 easily imply that for monotone functions the converse of Theorem 1.21 holds.

Another application of Proposition 4.5 gives a general upper bound for the total influence for monotone functions.

Proposition 4.6 *If $f: \Omega_n \to \{-1, 1\}$ or $\{0, 1\}$ is monotone, then*

$$\mathbf{I}(f) \leq \sqrt{n}.$$

Proof If the image is $\{-1, 1\}$, then by Proposition 4.5, we have

$$\mathbf{I}(f) = \sum_{k=1}^{n} \mathbf{I}_k(f) = \sum_{k=1}^{n} \hat{f}(\{k\}).$$

By the Cauchy–Schwarz inequality, this is at most $(\sum_{k=1}^{n} \hat{f}^2(\{k\}))^{1/2} \sqrt{n}$. By Parseval's formula, the first term is at most 1 and we are done. If the image is $\{0, 1\}$, this argument can easily be modified or we can deduce the result from the first case because the total influence of the corresponding ± 1-valued function is the same.   □

Remark Proposition 4.6 with some universal c on the right-hand side follows (for odd n) from Exercise 3.5 showing that the Majority function has the largest influence together with the known influences for Majority. However, the preceding argument yields a more direct proof of the $\sqrt{n}$ upper bound.

4.6 Exercises

4.1 Prove the discrete Poincaré inequality, Theorem 1.13, using the spectrum.

4.2 Compute the Fourier coefficients for the indicator function that there are all 1's.

4.3 Show that all even-size Fourier coefficients for the Majority function are 0. Can you extend this result to a broader class of Boolean functions?

4.4 For the Majority function $\mathbf{MAJ}_n$, find the limit (as the number of voters n goes to infinity) of the following quantity (total weight of the level-3 Fourier coefficients):

$$E_{\mathbf{MAJ}_n}(3) := \sum_{|S|=3} \widehat{\mathbf{MAJ}}_n(S)^2 .$$

4.5 Let $\{f_n\}$ be a sequence of Boolean functions that is noise sensitive and $\{g_n\}$ be a sequence of Boolean functions that is noise stable. Show that $\{f_n\}$ and $\{g_n\}$ are asymptotically uncorrelated.

4.6 Let $\{A_n\}$ be the event that the first bit is 1 and that the number of 1's in the last $n - 1$ bits is even. Show directly that this sequence is neither noise stable nor noise sensitive. Describe the Fourier spectrum as precisely as possible. You should see both a nontrivial amount of spectrum "near 0" and a nontrivial amount of spectrum "far from 0."

4.7 (Another equivalent definition of noise sensitivity)   Assume that $\{A_n\}$ is a noise-sensitive sequence. (This of course means that the indicator functions of these events is a noise-sensitive sequence.)

(a) Show for each $\epsilon > 0$, we have that $\mathbb{P}[\omega_\epsilon \in A_n \,|\, \omega] - \mathbb{P}[A_n]$ approaches 0 in probability. *Hint:* Use the Fourier representation.
(b) Can you show the implication in (a) without using the Fourier representation?
(c) Discuss whether this implication is surprising.
(d) Show that the condition in part (a) implies that the sequence is noise sensitive directly without using the Fourier representation.

4.8 How does the spectrum of a generic Boolean function look? Use this to give an alternative argument to the answer to the question asked in Exercise 1.14.

4.9 *(Open exercise)*. For Boolean functions, can the energy spectrum take ANY (reasonable) shape or are there restrictions?

For the next exercises, we introduce a functional that measures the stability of Boolean functions. For any Boolean function $f : \Omega_n \to \{-1, 1\}$, *let*

$$\mathbb{S}_f : \epsilon \mapsto \mathbb{P}[f(\omega) \neq f(\omega_\epsilon)].$$

Obviously, the smaller $\mathbb{S}_f$ *is, the more stable* f *is.*

4.10 Express the functional $\mathbb{S}_f$ in terms of the Fourier expansion of f.

A **balanced** *Boolean function takes its two possible values each with probability* $1/2$.

4.11 Among balanced Boolean functions, does there exist some function f^* that is "stablest" in the sense that for any balanced Boolean function f and any $\epsilon > 0$,

$$\mathbb{S}_{f^*}(\epsilon) \leq \mathbb{S}_f(\epsilon)?$$

If yes, describe the set of these extremal functions and prove that these are the only ones.

4.12 Show that for any Boolean function $\mathbb{S}_f'(0) = \mathbf{I}(f)/2$ and $\mathbb{S}_f'(1) = E_f(1)/2$ (recall Definition 4.1). Observe that for monotone functions, the last expression is $\sum_k \mathbf{I}_k(f)^2/2$.

Note that this yields an interesting heuristic why Theorem 1.21 is true in the monotone case. If $\sum_k \mathbf{I}_k(f_n)^2$ goes to 0, then $\mathbb{S}_f'(1)$ goes to 0. This suggests that $\mathbb{S}_f(1 - \epsilon)$ should be close to $\mathbb{S}_f(1)$, which corresponds to the completely uncorrelated case. Using the concavity of $\mathbb{S}_f$ and this picture, one can see in another way why noise sensitivity implies that $\sum_k \mathbf{I}_k(f_n)^2$ goes to 0.

4.13 This problem explores the *asymptotic shape* of the energy spectrum for **MAJ**$_n$.

(a) Show that for all $\epsilon \geq 0$,

$$\lim_{n \to \infty} \mathbb{S}_{\mathbf{MAJ}_n}(\epsilon) = \frac{1}{2} - \frac{\arcsin(1 - \epsilon)}{\pi} = \frac{\arccos(1 - \epsilon)}{\pi}.$$

Hint: The relevant limit is easily expressed as the probability that a certain two-dimensional Gaussian variable (with a particular correlation structure) falls in a certain area of the plane. One can write down the corresponding density function and therefore this probability as an explicit integral, but this integral does not seem so easy to evaluate. However, this Gaussian probability can be computed directly by representing the joint distribution in terms of two independent Gaussians.

Note that the above limit immediately implies that for $f_n = \mathbf{MAJ}_n$,

$$\lim_{n \to \infty} \mathbb{E}(f_n(\omega)f_n(\omega_\epsilon)) = \frac{2\arcsin(1-\epsilon)}{\pi}.$$

(b) Deduce from (a) and the Taylor expansion for $\arcsin(x)$ the limiting value, as $n \to \infty$, of $\mathbf{E_{MAJ_n}}(k) = \sum_{|S|=k} \widehat{\mathbf{MAJ}}_n(S)^2$ for all $k \geq 1$. Check that the answer is consistent with the values obtained earlier for $k = 1$ and $k = 3$ (see Exercise 4.4).

5

Hypercontractivity and its applications

In this chapter, we prove the main theorems about influences stated in Chapter 1. The proofs rely on techniques imported from harmonic analysis, in particular *hypercontractivity*, and as we see later in this chapter and in Chapter 7, these types of argument extend to other contexts that will be of interest to us such as noise sensitivity and sub-Gaussian fluctuations.

5.1 Heuristics of proofs

All the following proofs, which will be based on *hypercontractivity*, have more or less the same flavor, so we first sketch in the particular case of Theorem 1.14 the overall scheme of proof.

Recall that we want to prove that there exists a universal constant $c > 0$ such that for any function $f : \Omega_n \to \{0, 1\}$, one of its variables has influence at least $c \frac{\log n \operatorname{Var}(f)}{n}$.

Let f be a Boolean function. Suppose all its influences $\mathbf{I}_k(f)$ are "small" (this needs to be quantified). This means that $\nabla_k f$ must have small support. Using intuition from the Weyl–Heisenberg uncertainty principle, $\widehat{\nabla_k f}$ should then be quite spread out in the sense that most of its spectral mass is concentrated on high frequencies.

This intuition, which is still vague at this point, suggests that having small influences pushes the spectrum of $\nabla_k f$ toward high frequencies. Now, summing up as we did in Section 4.4, but restricting ourselves to frequencies S of size smaller than some large (well-chosen) M with $1 \ll M \ll n$, we easily obtain

$$\sum_{0<|S|<M} \hat{f}(S)^2 \leq 4 \sum_{0<|S|<M} |S| \hat{f}(S)^2$$

$$= \sum_k \sum_{0<|S|<M} \widehat{\nabla_k f}(S)^2$$

$$\text{``}\ll\text{''} \sum_k \|\nabla_k f\|_2^2$$

$$= \mathbf{I}(f), \tag{5.1}$$

where, in the third line, we use the informal statement that $\widehat{\nabla_k f}$ should be supported on high frequencies if f has small influences. Now recall (or observe) that

$$\sum_{|S|>0} \hat{f}(S)^2 = \mathrm{Var}(f).$$

Therefore, if we are in the case where a positive fraction of the Fourier mass of f is concentrated below M, then (5.1) says that $\mathbf{I}(f)$ is much larger than $\mathrm{Var}(f)$. In particular, at least one of the influences has to be "large." If, on the other hand, we are in the case where most of the spectral mass of f is supported on frequencies of size higher than M, then we also can conclude that $\mathbf{I}(f)$ is large by using the formula

$$\mathbf{I}(f) = 4 \sum_S |S| \hat{f}(S)^2.$$

Remark These heuristics suggest that there is a subtle balance between $\sum_k \mathbf{I}_k(f) = \mathbf{I}(f)$ and $\sup_k \mathbf{I}_k(f)$. Namely, if influences are all small (i.e., $\|\cdot\|_\infty$ is small), then their sum on the other hand has to be "large." The correct balance is exactly quantified by Theorem 1.16.

It now remains to convert this sketch into a proof. The main difficulty in the program we have outlined is to obtain quantitative spectral information on functions with values in $\{-1, 0, 1\}$ knowing that they have small support. This is where hypercontractivity makes its entrance (KKL88).

5.2 About hypercontractivity

First, let us discuss what we mean by hypercontractivity. Let $(K_t)_{t\geq 0}$ be the heat kernel for the Ornstein–Ulhenbeck process on $\mathbb{R}^n$. Hypercontractivity is a statement that quantifies how functions are regularized under this (Ornstein–Ulhenbeck) heat flow. The statement, which can be attributed to a number of authors, is simple:

Theorem 5.1 (Hypercontractivity)   *Consider $\mathbb{R}^n$ with standard Gaussian measure. If $1 < q < 2$, there is some $t = t(q) > 0$ (which does not depend on the dimension n) such that for any $f \in L^q(\mathbb{R}^n)$,*

$$\|K_t * f\|_2 \leq \|f\|_q.$$

The dependence $t = t(q)$ can be made explicit. (See Theorem 5.10 later in this chapter.) Hypercontractivity is thus a regularization statement: if we start with some initial "rough" L^q function f outside of L^2 and wait long enough $(t(q))$ under the (Ornstein–Uhlenbeck) heat flow, we end up in L^2 with good control on the functions' L^2 norm.

This concept has an interesting history, as is nicely explained in O'Donnell's book (see (O'D)). It was invented by Nelson in (Nel66) when he needed regularization estimates on free fields (which are the building blocks of quantum field theory) to apply these in "constructive field theory." It was then generalized by Gross in his elaboration of logarithmic Sobolev inequalities (Gro75), an important tool in analysis. Since then, hypercontractivity has become intimately related to these log-Sobolev inequalities and thus has many applications in the theory of semigroups, mixing of Markov chains, and other topics. In fact Gross obtained his hypercontractivity theorem for the Gaussian measure by first proving it in the Boolean setting and then passing to the limit via the central limit theorem. Gross was not aware of earlier work by Bonami (B70) that settled the Boolean case. Finally, let us mention that the hypercontractivity theorem has also been attributed in the past to Beckner (B75), even though this seems less convincing to us. For all these reasons, instead of calling this theorem the Nelson–Bonami–Gross–Beckner theorem, we will simply say the hypercontractivity theorem.

We now state the result in the case that concerns us, namely the hypercube. For any $\rho \in [0, 1]$, let T_ρ be the following **noise operator** on the set of functions on the hypercube: recall from Chapter 1 that if $\omega \in \Omega_n$, we denote by ω_ϵ an ϵ-noised configuration of ω. For any $f : \Omega_n \to \mathbb{R}$, we define $T_\rho f : \omega \mapsto \mathbb{E}[f(\omega_{1-\rho}) \mid \omega]$. This noise operator acts in a very simple way on the Fourier coefficients, as readers can check:

$$T_\rho : f = \sum_S \hat{f}(S)\chi_S \mapsto \sum_S \rho^{|S|}\hat{f}(S)\chi_S.$$

We have the following analog of Theorem 5.1.

Theorem 5.2 (Hypercontractivity)   *For any $f : \Omega_n \to \mathbb{R}$ and any $\rho \in [0, 1]$,*

$$\|T_\rho f\|_2 \leq \|f\|_{1+\rho^2}.$$

The analogy with the classical Theorem 5.1 is clear: the (Ornstein–Ulhenbeck) heat flow is replaced here by random walk on the hypercube. In order not to disrupt our exposition, the proof of Theorem 5.2 can be found in Section 5.6.

Remark The term *hypercontractive* refers here to the fact that the operator that maps L^q into L^2 (with $q < 2$) is a contraction.

Before going into the detailed proof of Theorem 1.14, let us see why Theorem 5.2 provides the type of spectral information we need. In the sketch in Section 5.1, we assumed that all influences were small. This assumption can be written as

$$\mathbf{I}_k(f) = \|\nabla_k f\|_1 = \|\nabla_k f\|_2^2 \ll 1,$$

for any $k \in [n]$. Now if we apply the hypercontractive estimate to these functions $\nabla_k f$ for some fixed $0 < \rho < 1$, we obtain

$$\|T_\rho(\nabla_k f)\|_2 \le \|\nabla_k f\|_{1+\rho^2} = \|\nabla_k f\|_2^{2/(1+\rho^2)} \ll \|\nabla_k f\|_2 \qquad (5.2)$$

where, for the equality, we used once again that $\nabla_k f \in \{-1, 0, 1\}$. After squaring, this gives on the Fourier side

$$\sum_S \rho^{2|S|} \widehat{\nabla_k f}(S)^2 \ll \sum_S \widehat{\nabla_k f}(S)^2,$$

showing that (under the assumption that $\mathbf{I}_k(f)$ is small) the spectrum of $\nabla_k f$ is indeed concentrated mostly on high frequencies.

Remarks Theorem 5.2 in fact tells us that any function with small support has its frequencies concentrated on large sets as follows. It is easy to see that given any $p < 2$, if a function h on a probability space has very small support, then its L_p norm is much smaller than its L_2 norm. Using Theorem 5.2, we would then have for such a function

$$\|T_\rho(h)\|_2 \le \|h\|_{1+\rho^2} \ll \|h\|_2,$$

yielding

$$\sum_S \rho^{2|S|} \widehat{h}(S)^2 \ll \sum_S \widehat{h}(S)^2,$$

which can occur only if h has its frequencies concentrated on large sets. From this point of view, we also see directly that, under the small influence assumption, the first term in (5.2) is of much smaller order than the last term in (5.2).

5.3 Proof of the KKL theorems on the influences of Boolean functions

We prove first Theorem 1.14, and then Theorem 1.16. In fact, one can recover Theorem 1.14 directly from Theorem 1.16; see Exercise 5.1. Nevertheless, because proof of Theorem 1.14 is slightly simpler, we start with this result.

5.3.1 Proof of Theorem 1.14

Let $f\colon \Omega_n \to \{0,1\}$. Recall that we want to show that there is some $k \in [n]$ such that

$$\mathbf{I}_k(f) \geq c\operatorname{Var}(f)\frac{\log n}{n}, \tag{5.3}$$

for some universal constant $c > 0$.

We divide the analysis into the following two cases.

Case 1

Suppose that there is some $k \in [n]$ such that $\mathbf{I}_k(f) \geq n^{-3/4}\operatorname{Var}(f)$. Then the bound 5.3 is clearly satisfied for a small enough $c > 0$.

Case 2

Now, if f does not belong to the first case, then for all $k \in [n]$,

$$\mathbf{I}_k(f) = \|\nabla_k f\|_2^2 \leq \operatorname{Var}(f)n^{-3/4}. \tag{5.4}$$

Following the heuristics in Section 5.1, we will show that when (5.4) holds for all k, most of the Fourier spectrum of f is supported on high frequencies. Let $M \geq 1$, with its value to be chosen later. We wish to bound from above the bottom part (up to M) of the Fourier spectrum of f:

$$\sum_{1 \leq |S| \leq M} \hat{f}(S)^2 \leq \sum_{1 \leq |S| \leq M} |S| \hat{f}(S)^2$$

$$\leq 2^{2M} \sum_{|S| \geq 1} (1/2)^{2|S|} |S| \hat{f}(S)^2$$

$$= \frac{1}{4} 2^{2M} \sum_k \|T_{1/2}(\nabla_k f)\|_2^2,$$

(see Section 4.4.) Now by applying hypercontractivity (Theorem 5.2) with $\rho = 1/2$ to the above sum, we obtain

$$\sum_{1 \leq |S| \leq M} \hat{f}(S)^2 \leq \frac{1}{4} 2^{2M} \sum_k \|\nabla_k f\|_{5/4}^2$$

$$\leq 2^{2M} \sum_k \mathbf{I}_k(f)^{8/5}$$

$$\leq 2^{2M} n \operatorname{Var}(f)^{8/5} n^{\frac{-3}{4} \cdot \frac{8}{5}}$$

$$\leq 2^{2M} n^{-1/5} \operatorname{Var}(f),$$

where we used assumption (5.4) and the obvious fact that $\operatorname{Var}(f)^{8/5} \leq \operatorname{Var}(f)$ (recall $\operatorname{Var}(f) \leq 1$ since f is Boolean). Now with $M := \lfloor \frac{1}{20} \log_2 n \rfloor$, this gives

$$\sum_{1 \leq |S| \leq \frac{1}{20} \log_2 n} \hat{f}(S)^2 \leq n^{1/10 - 1/5} \operatorname{Var}(f) = n^{-1/10} \operatorname{Var}(f).$$

This shows that under assumption (5.4), most of the Fourier spectrum is concentrated above $\Omega(\log n)$. We are now ready to conclude

$$\sup_k \mathbf{I}_k(f) \geq \frac{\sum_k \mathbf{I}_k(f)}{n} = \frac{4 \sum_{|S| \geq 1} |S| \hat{f}(S)^2}{n}$$

$$\geq \frac{1}{n} [\sum_{|S| > M} |S| \hat{f}(S)^2]$$

$$\geq \frac{M}{n} [\sum_{|S| > M} \hat{f}(S)^2]$$

$$= \frac{M}{n} [\operatorname{Var}(f) - \sum_{1 \leq |S| \leq M} \hat{f}(S)^2]$$

$$\geq \frac{M}{n} \operatorname{Var}(f) [1 - n^{-1/10}]$$

$$\geq c_1 \operatorname{Var}(f) \frac{\log n}{n},$$

with $c_1 = \frac{1}{20 \log 2}(1 - 2^{-1/10})$. By combining with the constant given in case 1, this completes the proof.   □

Remark We did not try here to optimize the proof to find the best possible universal constant $c > 0$. Note, though, that even without optimizing at all, the constant we obtain is not that bad.

5.3.2 Proof of Theorem 1.16

We now proceed to the proof of the stronger result, Theorem 1.16, which states that there is a universal constant $c > 0$ such that for any $f : \Omega_n \to \{0, 1\}$,

$$\mathbf{I}(f) = \|\mathbf{Inf}(f)\|_1 \geq c \operatorname{Var}(f) \log \frac{1}{\|\mathbf{Inf}(f)\|_\infty}.$$

The strategy is very similar to that for Theorem 1.14. Let $f : \Omega_n \to \{0, 1\}$ and let $\delta := \|\mathbf{Inf}(f)\|_\infty = \sup_k \mathbf{I}_k(f)$. Assume for the moment that $\delta \leq 1/1000$. As in the proof of Theorem 1.14, we start by bounding the bottom part of the spectrum up to some integer M (whose value will be fixed later). Exactly, as before, one has

$$\sum_{1 \leq |S| \leq M} \hat{f}(S)^2 \leq 2^{2M} \sum_k \mathbf{I}_k(f)^{8/5}$$

$$\leq 2^{2M} \delta^{3/5} \sum_k \mathbf{I}_k(f) = 2^{2M} \delta^{3/5} \mathbf{I}(f).$$

Now,

$$\operatorname{Var}(f) = \sum_{|S| \geq 1} \hat{f}(S)^2 \leq \sum_{1 \leq |S| \leq M} \hat{f}(S)^2 + \frac{1}{M} \sum_{|S| > M} |S| \hat{f}(S)^2$$

$$\leq \left[2^{2M} \delta^{3/5} + \frac{1}{M} \right] \mathbf{I}(f).$$

Choose $M := \frac{3}{10} \log_2(\frac{1}{\delta}) - \frac{1}{2} \log_2 \log_2(\frac{1}{\delta})$. Since $\delta < 1/1000$, it is easy to check that $M \geq \frac{1}{10} \log_2(1/\delta)$ which leads to

$$\operatorname{Var}(f) \leq \left[\frac{1}{\log_2(1/\delta)} + \frac{10}{\log_2(1/\delta)} \right] \mathbf{I}(f),$$

$$(5.5)$$

which gives

$$\mathbf{I}(f) = \|\mathbf{Inf}(f)\|_1 \geq \frac{1}{11\log 2} \operatorname{Var}(f) \log \frac{1}{\|\mathbf{Inf}(f)\|_\infty}.$$

This gives the result for $\delta \leq 1/1000$.

Next the discrete Poincaré inequality, which says that $\mathbf{I}(f) \geq \operatorname{Var}(f)$, tells us that the claim is true for $\delta \geq 1/1000$ if we take c to be $1/\log 1000$. Because this is larger than $\frac{1}{11\log 2}$, we obtain the result with the constant $c = \frac{1}{11\log 2}$. $\square$

5.4 KKL away from the uniform measure

In Chapter 3 (on sharp thresholds), we needed an extension of the KKL theorems to the p-biased measures $\mathbb{P}_p = (p\delta_1 + (1-p)\delta_{-1})^{\otimes n}$. These extensions are respectively Theorems 3.3 and 3.4.

To prove these theorems a natural first idea would be to extend the hypercontractive estimate (Theorem 5.2) to these p-biased measures $\mathbb{P}_p$. This extension of the hypercontractivity theorem is possible, but it turns out that the control it gives gets worse near the edges (p close to 0 or 1). This is problematic because in both Theorems 3.3 and 3.4, we need bounds that are uniform in $p \in [0, 1]$.

Hence, we need a different way to extend the KKL theorems. A nice approach was provided in (BKK+92), where the authors prove the following general theorem.

Theorem 5.3 (BKK+92) *There exists a universal $c > 0$ such that for any measurable function $f: [0,1]^n \to \{0,1\}$, there exists a variable k such that*

$$\mathbf{I}_k(f) \geq c \operatorname{Var}(f) \frac{\log n}{n}.$$

Here the "continuous" hypercube is endowed with the uniform (Lebesgue) measure and for any $k \in [n]$, $\mathbf{I}_k(f)$ denotes the probability that f is not almost surely constant on the fiber given by $(x_i)_{i \neq k}$.

In other words,

$$\mathbf{I}_k(f) = \mathbb{P}\left[\operatorname{Var}(f(x_1, \ldots, x_n)) \mid x_i, i \neq k \right) > 0 \right].$$

It is clear how to obtain Theorem 3.3 from Theorem 5.3. If $p \in [0,1]$ and $f: \Omega_n \to \{0,1\}$, consider $\bar{f}_p: [0,1]^n \to \{0,1\}$ defined by

$$\bar{f}_p(x_1, \ldots, x_n) = f((1_{x_i < p} - 1_{x_i \geq p})_{i \in [n]}).$$

Friedgut (Fri04) noticed the converse, that one can recover Theorem 5.3 from Theorem 3.3. The first idea is to use a symmetrization argument in such a way that the problem reduces to the case of monotone functions. Then, the main idea is to approximate the uniform measure on $[0, 1]$ by the dyadic random variable

$$X_M : (x_1, \ldots, x_M) \in \{-1, 1\}^M \mapsto \sum_{m=1}^{M} \frac{x_m + 1}{2} 2^{-m}.$$

One can then approximate $f : [0, 1]^n \to \{0, 1\}$ by the Boolean function $\tilde{f}_M$ defined on $\{-1, 1\}^{M \times n}$ by

$$\tilde{f}_M(x_1^1, \ldots, x_M^1, \ldots, x_1^n, \ldots, x_M^n) := f(X_M^1, \ldots, X_M^n).$$

Still this proof requires two technical steps: a monotonization procedure and an "approximation" step (going from f to $\tilde{f}_M$). Because in our applications to sharp thresholds we used Theorems 3.3 and 3.4 only in the case of monotone functions, for the sake of simplicity we do not present the monotonization procedure here.

Furthermore, for our specific needs (the applications in Chapter 3), we do not need to deal with the approximation part either. The reason is that for any Boolean function f, the function $p \mapsto \mathbf{I}_k^p(f)$ is continuous. Hence it is enough to obtain uniform bounds on $\mathbf{I}_k^p(f)$ for dyadic values of p (i.e., $p \in \{m2^{-M}\} \cap [0, 1]$).

See Exercise 5.4 for the proof of Theorems 3.3 and 3.4.

Remark We mentioned previously that generalizing hypercontractivity would not allow us to obtain uniform bounds (with p taking any value in $[0, 1]$) on the influences. It should be noted though that Talagrand (Tal94) obtained results similar to Theorems 3.3 and 3.4 by somehow generalizing hypercontractivity, but along a different line. Finally, let us point out that both Talagrand (see (Tal94)) and Friedgut and Kalai (FK96) obtain sharper versions of Theorems 3.3 and 3.4 where the constant $c = c_p$ in fact improves (i.e., blows up) near the edges.

We state the result due to Talagrand. It is this result that also implies Theorem 3.8.

Theorem 5.4 (Tal94)   *There exists $c > 0$ such that for all Boolean functions and all p, we have*

$$\mathrm{Var}_p(f) \leq cp(1 - p) \log\left(\frac{1}{p(1 - p)}\right) \sum_i \frac{I_i^p(f)}{\log(1/(p(1 - p)I_i^p(f)))}.$$

5.5 The noise sensitivity theorem

In this section, we prove the milestone Theorem 1.21 from (BKS99) under an additional assumption (which will hold whenever the theorem is applied). Before recalling what the statement is, let us define the following functional on Boolean functions.

Definition 5.5 For any $f \colon \Omega_n \to \{0,1\}$, let

$$\mathbf{H}(f) := \sum_k \mathbf{I}_k(f)^2 = \|\mathbf{Inf}(f)\|_2^2.$$

Recall the Benjamini–Kalai–Schramm theorem, Theorem 1.21, states that, for a sequence of Boolean functions $f_n \colon \Omega_{m_n} \to \{0,1\}$, if

$$\lim_{n\to\infty} \mathbf{H}(f_n) = \sum_{k=1}^{m_n} \mathbf{I}_k(f)^2 = 0,$$

then $\{f_n\}_n$ is noise sensitive.

We will in fact prove this theorem only under a stronger condition, namely that $\mathbf{H}(f_n) \le (m_n)^{-\delta}$ for some exponent $\delta > 0$. Without this assumption of "polynomial decay" on $\mathbf{H}(f_n)$, the proof is more technical and relies on estimates obtained by Talagrand. See the discussion immediately following the proof of Proposition 5.6. For our application to the *noise sensitivity of percolation* (see Chapter 6), this stronger assumption will be satisfied and hence we stick to this simpler case in this book.

The assumption of polynomial decay in fact enables us to prove the following more quantitative result.

Proposition 5.6 (BKS99) *For any $\delta > 0$, there exists a constant $M = M(\delta) > 0$ such that if $f_n \colon \Omega_{m_n} \to \{0,1\}$ is any sequence of Boolean functions satisfying*

$$\mathbf{H}(f_n) \le (m_n)^{-\delta}, \tag{5.6}$$

then

$$\sum_{1 \le |S| \le M \log(m_n)} \widehat{f_n}(S)^2 \to 0.$$

Using Proposition 4.2, this proposition obviously implies Theorem 1.21 when $\mathbf{H}(f_n)$ decays as assumed. Furthermore, this gives quantitative "logarithmic" control on the noise sensitivity of such functions.

Proof The strategy will be very similar to that used for the KKL theorems (even though the goal is very different). The main difference here is that the regularization term ρ used in the hypercontractive estimate must be chosen in a more delicate way than in the proofs of the KKL results (where we simply took $\rho = 1/2$).

Let $M > 0$ be a constant whose value will be chosen later:

$$\sum_{1 \le |S| \le M \log(m_n)} \widehat{f_n}(S)^2 \le 4 \sum_{1 \le |S| \le M \log(m_n)} |S| \widehat{f_n}(S)^2 = \sum_k \sum_{1 \le |S| \le M \log(m_n)} \widehat{\nabla_k f_n}(S)^2$$

$$\le \sum_k \left(\frac{1}{\rho^2}\right)^{M \log(m_n)} \|T_\rho(\nabla_k f_n)\|_2^2$$

$$\le \sum_k \left(\frac{1}{\rho^2}\right)^{M \log(m_n)} \|\nabla_k f_n\|_{1+\rho^2}^2$$

by Theorem 5.2.

Now, since f_n is Boolean, one has $\|\nabla_k f_n\|_{1+\rho^2} = \|\nabla_k f_n\|_2^{2/(1+\rho^2)}$, hence

$$\sum_{0 < |S| < M \log(m_n)} \widehat{f_n}(S)^2 \le \rho^{-2M \log(m_n)} \sum_k \|\nabla_k f_n\|_2^{4/(1+\rho^2)} = \rho^{-2M \log(m_n)} \sum_k \mathbf{I}_k(f_n)^{2/(1+\rho^2)}$$

$$\le \rho^{-2M \log(m_n)} (m_n)^{\rho^2/(1+\rho^2)} \left(\sum_k \mathbf{I}_k(f_n)^2\right)^{\frac{1}{1+\rho^2}} \text{ (by Hölder)}$$

$$= \rho^{-2M \log(m_n)} (m_n)^{\rho^2/(1+\rho^2)} \mathbf{H}(f_n)^{\frac{1}{1+\rho^2}}$$

$$\le \rho^{-2M \log(m_n)} (m_n)^{\frac{\rho^2-\delta}{1+\rho^2}}.$$

By choosing $\rho \in (0, 1)$ close enough to 0, and then by choosing $M = M(\delta)$ small enough, we obtain the desired logarithmic noise sensitivity. $\square$

Recall that Theorem 1.21 is true independently of the speed of convergence of $\mathbf{H}(f_n) = \sum_k \mathbf{I}_k(f_n)^2$ to 0. The proof of this general result is a bit more involved than the argument we gave here for the special case when the convergence of $\mathbf{H}(f_n)$ to 0 is "at least inverse polynomial." We now indicate how the proof of Theorem 1.21 works in the general case. The key lemma needed is the following.

Lemma 5.7 (BKS99) *There exist absolute constants C_k such that for any monotone Boolean function f and for any $k \ge 2$, one has*

$$\sum_{|S|=k} \hat{f}(S)^2 \le C_k \mathbf{H}(f)(-\log \mathbf{H}(f))^{k-1}.$$

This lemma mimics a result from Talagrand's (Tal96). Indeed, Proposition 2.3 in (Tal96) can be translated into Lemma 5.7 with $k = 2$. Lemma 5.7 obviously implies Theorem 1.21 in the monotone case, while the general case can be deduced by a monotonization procedure; this latter procedure is detailed in Lemma 2.7 in (BKS99). It is worth pointing out that hyper-contractivity is used in the proof of Lemma 5.7.

Theorem 1.21, as stated, is clearly a qualitative result. One would like to have a quantitative version; this was obtained in (BKS99) under certain conditions. The following result from (KK13) yields a general quantitative result.

Theorem 5.8 (KK13) *There exists $c > 0$ such that for any Boolean function f mapping into $\{0, 1\}$ and any $\epsilon > 0$, it is the case that*

$$\mathbb{E}[f(\omega)f(\omega_\epsilon)] - \mathbb{E}[f(\omega)]^2 \leq 20(\mathbf{H}(f))^{c\epsilon}.$$

We end this section by mentioning that (BKS99) states that there is a version of Theorem 1.21 for biased measures that is proved along the same lines. There is also an approach for this given in (ABGM13) that involves reduction to the uniform case. Lastly, there is a quantitative version in the biased case akin to Theorem 5.8 above; see again (KK13).

5.6 Proof of hypercontractivity

We will prove Theorem 5.2 from first principles.

Before starting the proof, observe that for $\rho = 0$ (where 0^0 is defined to be 1), this simply reduces to $|\int f| \leq \int |f|$.

5.6.1 Tensorization

First, we show that it is sufficient, via a tensorization procedure, that the result holds for $n = 1$ in order for us to conclude by induction the result for all n.

The key step of the argument is the following lemma.

Lemma 5.9 *Let $q \geq p \geq 1$, $(\Omega_1, \mu_1), (\Omega_2, \mu_2)$ be two finite probability spaces, $K_i \colon \Omega_i \times \Omega_i \to \mathbb{R}$ and assume that for $i = 1, 2$ and for all $f \colon \Omega_i \to \mathbb{R}$*

$$\|T_i(f)\|_{L_q(\Omega_i, \mu_i)} \leq \|f\|_{L_p(\Omega_i, \mu_i)},$$

where $T_i(f)(x) := \int_{\Omega_i} f(y)K_i(x, y)d\mu_i(y)$. Then for all $f \colon \Omega_1 \times \Omega_2 \to \mathbb{R}$

$$\|T_1 \otimes T_2(f)\|_{L_q((\Omega_1, \mu_1) \times (\Omega_2, \mu_2))} \leq \|f\|_{L_p((\Omega_1, \mu_1) \times (\Omega_2, \mu_2))},$$

where $T_1 \otimes T_2(f)(x_1, x_2) := \int_{\Omega_1 \times \Omega_2} f(y_1, y_2)K_1(x_1, y_1)K_2(x_2, y_2)d\mu_1(y_1)$ $\times d\mu_2(y_2)$.

Proof First recall Minkowski's inequality for integrals, which states that, for $g \geq 0$ and $r \in [1, \infty)$, we have

$$\left(\int \left(\int g(x,y) d\nu(y) \right)^r d\mu(x) \right)^{1/r} \leq \int \left(\int g(x,y)^r d\mu(x) \right)^{1/r} d\nu(y).$$

(Note that when ν consists of two point masses each of size 1, then this reduces to the usual Minkowski inequality.)

Think of T_1 acting on functions of both variables by leaving the second variable untouched and analogously for T_2. It is then easy to check that $T_1 \otimes T_2 = T_1 \circ T_2$. Fix now $f : \Omega_1 \times \Omega_2 \to \mathbb{R}$. By thinking of x_2 as fixed, our assumption on T_1 yields

$$\|T_1 \otimes T_2(f)\|^q_{L_q((\Omega_1,\mu_1) \times (\Omega_2,\mu_2))} \leq \int_{\Omega_2} \left(\int_{\Omega_1} |T_2(f)|^p d\mu_1(x_1) \right)^{q/p} d\mu_2(x_2).$$

(It might be helpful here to think of $T_2(f)(x_1, x_2)$ as a function $g^{x_2}(x_1)$ where x_2 is fixed, and it is this function of x_1 to which we apply our assumption concerning T_1.)

Applying Minkowski's integral inequality to $|T_2(f)|^p$ with $r = q/p$, this in turn is at most

$$\left[\int_{\Omega_1} \left(\int_{\Omega_2} |T_2(f)|^q d\mu_2(x_2) \right)^{p/q} d\mu_1(x_1) \right]^{q/p}.$$

Fixing now the x_1 variable and applying our assumption on T_2 to the function $f(x_1, \cdot)$ shows that this is at most $\|f\|^q_{L_p((\Omega_1,\mu_1) \times (\Omega_2,\mu_2))}$, as desired. $\square$

The next key observation, easily obtained by expanding and interchanging of summation, is that our operator T_ρ acting on functions on Ω_n corresponds to an operator of the type dealt with in Lemma 5.9 with $K(x,y)$ being

$$\sum_{S \subseteq \{1,\dots,n\}} \rho^{|S|} \chi_S(x) \chi_S(y).$$

In addition, it is easy to check that the function K for Ω_n is simply an n-fold product of the function for the $n = 1$ case.

Assuming the result for the case $n = 1$, Lemma 5.9 and the above observations allow us to conclude by induction the result for all n.

5.6.2 The n= 1 case

We now establish the case $n = 1$. We abbreviate T_ρ by T.

Since $f(x) = (f(-1) + f(1))/2 + (f(1) - f(-1))/2\, x$, we have $Tf(x) = (f(-1) + f(1))/2 + \rho(f(1) - f(-1))/2\, x$. Denoting $(f(-1) + f(1))/2$ by a and $(f(1) - f(-1))/2$ by b, it suffices to show that for all a and b, we have

$$(a^2 + \rho^2 b^2)^{(1+\rho^2)/2} \leq \frac{|a + b|^{1+\rho^2} + |a - b|^{1+\rho^2}}{2}.$$

Using $\rho \in [0, 1]$, the case $a = 0$ is immediate. For the case, $a \neq 0$, it is clear we can assume $a > 0$. Dividing both sides by $a^{1+\rho^2}$, we need to show that

$$(1 + \rho^2 y^2)^{(1+\rho^2)/2} \leq \frac{|1 + y|^{1+\rho^2} + |1 - y|^{1+\rho^2}}{2} \tag{5.7}$$

for all y and clearly it suffices to assume $y \geq 0$.

We first do the case that $y \in [0, 1)$. By the generalized binomial formula, the right-hand side of (5.7) is

$$\frac{1}{2}\left[\sum_{k=0}^{\infty}\binom{1+\rho^2}{k}y^k + \sum_{k=0}^{\infty}\binom{1+\rho^2}{k}(-y)^k\right] = \sum_{k=0}^{\infty}\binom{1+\rho^2}{2k}y^{2k}.$$

For the left-hand side of (5.7), we first note the following. For $0 < \lambda < 1$, a simple calculation shows that the function $g(x) = (1 + x)^\lambda - 1 - \lambda x$ has a negative derivative on $[0, \infty)$ and hence $g(x) \leq 0$ on $[0, \infty)$.

This yields that the left-hand side of (5.7) is at most

$$1 + \left(\frac{1+\rho^2}{2}\right)\rho^2 y^2,$$

which is precisely the first two terms of the right-hand side of (5.7). On the other hand, the binomial coefficients appearing in the other terms are non-negative, because in the numerator there are an even number of terms with the first two terms being positive and all the other terms being negative. This verifies the desired inequality for $y \in [0, 1)$.

The case $y = 1$ for (5.7) follows by continuity.

For $y > 1$, we let $z = 1/y$ and note, by multiplying both sides of (5.7) by $z^{1+\rho^2}$, that we need to show

$$(z^2 + \rho^2)^{(1+\rho^2)/2} \leq \frac{|1 + z|^{1+\rho^2} + |1 - z|^{1+\rho^2}}{2}. \tag{5.8}$$

Now, expanding $(1 - z^2)(1 - \rho^2)$, one sees that $z^2 + \rho^2 \leq 1 + z^2\rho^2$ and hence the desired inequality follows precisely from (5.7) for the case $y \in (0, 1)$

already proved. This completes the $n = 1$ case and thereby the proof of Theorem 5.2. □

5.7 Gaussian hypercontractivity

We finish this chapter with a slightly tangential topic. Historically, Gross's main motivation in (Gro75) was to obtain a hypercontractive inequality for the Gaussian measure and he proved hypercontractivity on the hypercube $\{-1, 1\}^n$ only as an intermediate step to obtaining the Gaussian case. In this section, we sketch how his argument goes.

Consider the Ornstein–Uhlenbeck process $(X_t^x)_{t \geq 0}$ in $\mathbb{R}^n$ starting at $x \in \mathbb{R}^n$ which is given by the following stochastic differential equation (SDE):

$$\begin{cases} X_0^x = x, \\ dX_t^x = -\frac{1}{2}X_t \, dt + dW_t, \end{cases} \tag{5.9}$$

where $(W_t)_{t \geq 0}$ is a standard n-dimensional Brownian motion. It is a classical fact that this SDE leads to a Feller semigroup whose invariant measure is the n-dimensional Gaussian measure

$$\gamma_n(dx) = \frac{1}{(2\pi)^{n/2}} e^{-\frac{|x|^2}{2}} \, dx.$$

Gross proved the following result.

Theorem 5.10 (Gross, hypercontractivity for the Gaussian variable, (Gro75)) *Let $K_t \colon L^2(\mathbb{R}^n, \gamma_n) \to L^2(\mathbb{R}^n, \gamma_n)$ be the (Ornstein-Uhlenbeck) heat-kernel operator:*

$$K_t(f)(x) := \mathbb{E}[f(X_t^x)].$$

Then, for any $t \geq 0$ and any $f \in L^2(\mathbb{R}^n, \gamma_n)$, one has

$$\|K_t(f)\|_{L^2(\mathbb{R}^n, \gamma_n)} \leq \|f\|_{L^{1+e^{-t}}(\mathbb{R}^n, \gamma_n)}.$$

Sketch of proof: Fix $f \in L^{1+e^{-t}}(\mathbb{R}^n, \gamma_n)$ and assume for the moment that it is continuous and compactly supported, that is, in $C_c(\mathbb{R}^n)$. The idea will be to approximate the Ornstein–Uhlenbeck process using simple random walks in $\frac{1}{M}\mathbb{Z}^n$. For any $t > 0$ if $X_0 \sim \gamma_n$ and $X_t = X_t^{X_0}$, it is easy to check that the stochastic process $(X_t)_{t \geq 0}$ is a Gaussian process with covariance $\text{Cov}[X_s, X_t] = e^{-|t-s|/2}I$. In particular, we can represent the vector (X_0, X_t) as follows: $X_t = e^{-t/2}X_0 + \sqrt{1 - e^{-t}}Z$, where $Z \sim \gamma_n$ is an independent Gaussian in $\mathbb{R}^n$. We can then write

$$\|K_t(f)\|_{L^2(\mathbb{R}^n, \gamma_n)}^2 = \int_{\mathbb{R}^n} \left(\mathbb{E}[f(e^{-t/2}x + \sqrt{1 - e^{-t}}Z)] \right)^2 \gamma_n(dx)$$

We consider for each $N \geq 1$, the following real-valued function $g_N \colon \Omega_{N \times n} := \{-1, 1\}^{N \times n} \to \mathbb{R}$:

$$g_N(x_1^1, \ldots, x_N^1, x_1^2, \ldots, x_N^2, \ldots, x_1^n, \ldots, x_N^n) := f\left(\frac{\sum x_i^1}{\sqrt{N}}, \frac{\sum x_i^2}{\sqrt{N}}, \ldots, \frac{\sum x_i^n}{\sqrt{N}}\right).$$

Since f is in $C_c(\mathbb{R}^n)$, it follows easily from the central limit theorem that

$$\|g_N\|_{L^{1+e^{-t}}(\Omega_{N \times n})} \to \|f\|_{L^{1+e^{-t}}(\mathbb{R}^n, \gamma_n)},$$

as $N \to \infty$. Furthermore, it is easy to check that for any $\rho \in [0, 1]$,

$$\|T_\rho(g_N)\|_2^2 \to \int_{\mathbb{R}^n} \left(\mathbb{E}[f(e^{-t/2}x + \sqrt{1 - e^{-t}}Z)]\right)^2 \gamma_n(dx),$$

as $N \to \infty$ with the correspondence $e^{-t/2} = \rho$. This follows from the fact that for each $k \in \{1, \ldots, n\}$

$$\mathrm{Cov}[X_0^k, X_t^k] = \lim_{N \to \infty} \mathrm{Cov}\left[\frac{\sum_i x_i^k}{\sqrt{N}}, \frac{\sum_i y_i^k}{\sqrt{N}}\right] = \rho,$$

where $\{y_i^k\}_{i \in [N]}$ is a noised version of $\{x_i^k\}_{i \in [N]}$ with $\mathbb{E}[x_i^k y_i^k] = \rho$. Now, using hypercontractivity for functions on the hypercube $\Omega_{N \times n}$ (Theorem 5.2), we have

$$\|T_\rho(g_N)\|_2 \leq \|g_N\|_{1+\rho^2}.$$

By passing to the limit $N \to \infty$, we thus obtain for any $f \in C_c(\mathbb{R}^n)$

$$\|K_t(f)\|_{L^2(\mathbb{R}^n, \gamma_n)} \leq \|f\|_{L^{1+e^{-t}}(\mathbb{R}^n, \gamma_n)}.$$

We conclude the proof by a classical density argument: given $f \in L^{1+e^{-t}} \cap L^2$, choose $f_n \in C_c(\mathbb{R}^n) \to f$ in $L^{1+e^{-t}}$ and in L^2 and note that $K_t(f_n) \to K_t(f)$ in L^2, since K_t is a contraction operator on L^2. $\square$

Remark A Markov operator is a contraction on every L_p space and the proof of Theorem 5.10 also yields the main inequality even if we only assume that $f \in L^{1+e^{-t}}(\mathbb{R}^n, \gamma_n)$.

Remark See the book in progress (Ch14) for a more direct proof in the Gaussian setting based on the logarithmic Sobolev inequality for the Gaussian measure. A proof of the logarithmic Sobolev inequality for the Gaussian measure can be found in (Feder69).

5.8 Exercises

5.1   Find a direct proof that Theorem 1.16 implies Theorem 1.14.

5.2   Is it true that the smaller the influences are, the more noise-sensitive the function is?

5.3   Prove that Theorem 5.3 indeed implies Theorem 3.3. *Hint:* Use the natural projection.

5.4   In this problem, we prove Theorems 3.3 and 3.4.

(a) Show that Theorem 3.4 implies Theorem 3.3 and hence one needs to prove only Theorem 3.4 (This is the basically the same as Exercise 5.1.)

(b) Show that it suffices to prove the result when $p = k/2^\ell$ for integers k and ℓ.

(c) Let $\Pi \colon \{0,1\}^\ell \to \{0, 1/2^\ell, \ldots, (2^\ell - 1)/2^\ell\}$ by $\Pi(x_1, \ldots, x_\ell) = \sum_{i=1}^{\ell} x_i/2^i$. Observe that if x is uniform, then $\Pi(x)$ is uniform on its range and that $\mathbb{P}(\Pi(x) \geq i/2^\ell) = (2^\ell - i)/2^\ell$ for $i = 0, 1, \ldots, 2^\ell - 1$.

(d) Define $g \colon \{0,1\}^\ell \to \{0,1\}$ by $g(x_1, \ldots, x_\ell) := I_{\{\Pi(x) \geq 1-p\}}$. Note that $\mathbb{P}(g(x) = 1) = p$.

(e) Define $\tilde{f} \colon \{0,1\}^{n\ell} \to \{0,1\}$ by

$$\tilde{f}(x_1^1, \ldots, x_\ell^1, x_1^2, \ldots, x_\ell^2, \ldots, x_1^n, \ldots, x_\ell^n)'$$
$$= f(g(x_1^1, \ldots, x_\ell^1), g(x_1^2, \ldots, x_\ell^2), \ldots, g(x_1^n, \ldots, x_\ell^n)).$$

Observe that $\tilde{f}$ (defined on $(\{0,1\}^{n\ell}, \pi_{1/2})$) and f (defined on $(\{0,1\}^n, \pi_p)$) have the same distribution and hence the same variance.

(f) Show (or observe) that $\mathbf{I}_{(r,j)}(\tilde{f}) \leq \mathbf{I}_r^p(f)$ for each $r = 1, \ldots, n$ and $j = 1, \ldots, \ell$. Deduce from Theorem 1.16 that

$$\sum_{r,j} \mathbf{I}_{(r,j)}(\tilde{f}) \geq c\,\mathrm{Var}(f)\log(1/\delta_p),$$

where $\delta_p := \max_i \mathbf{I}_i^p(f)$ and where c comes from Theorem 1.16.

(g) (Key step). Show that for each $r = 1, \ldots, n$ and $j = 1, \ldots, \ell$,

$$\mathbf{I}_{(r,j)}(\tilde{f}) \leq \mathbf{I}_r^p(f)/2^{j-1}.$$

(h) Combine parts (f) and (g) to complete the proof.

The next three exercises are tangential to the book but are interesting nonetheless.

5.5 Consider a real-valued function f defined on Ω_n all of whose nonzero Fourier coefficients have level at most k. Show, using Theorem 5.2, that for all $p \in (1,2)$

$$\mathbb{E}(f^2)^{\frac{1}{2}} \le (1/(p-1))^{\frac{k}{2}} \mathbb{E}(|f|^p)^{\frac{1}{p}}.$$

(When $k = 1$, this gives a version of the so-called Khinchin inequalities.)

5.6 With f as in Exercise 5.5, show that

$$\mathbb{E}(f^2)^{\frac{1}{2}} \le 2^{\frac{3k}{2}} \mathbb{E}(|f|).$$

Hint: First use the Cauchy–Schwarz inequality applied to the 3/2–moment of f to show that the ratio of the L^2 and L^1 norms of f is at most the third power of the ratio of the L^2 and $L^{\frac{3}{2}}$ norms of f. Then apply Theorem 5.2 as in Exercise 5.5.

5.7 Let $X_1, \ldots, X_n$ be i.i.d. ± 1 mean 0 random variables and let $Z = |\sum_i b_i X_i|$. Show using Fourier analysis, but without Theorem 5.2, that $\mathbb{E}(Z^2)^{\frac{1}{2}} \le 2^{\frac{1}{2}} \mathbb{E}(Z)$. The steps below give an outline of the argument.

Remarks

(1) This result is due to Szarek.
(2) This reduces to the result in Exercise 5.6 but with an improvement of the constant.
(3) The result can be seen to be sharp by taking $n = 2$ and $b_1 = b_2 = 1$.

(4) The argument below applies if the b_i take values in a Banach space and Z is replaced by the norm of the sum.

(a) Thinking of Z as a Boolean function f on the hypercube, show that the odd Fourier coefficients of f vanish.
(b) Letting $\overline{f}(\omega) := \sum_{i=1}^n f(\sigma_i(\omega))$, show that

$$\overline{f}(\omega) = \sum_{S \subseteq \{1,\ldots,n\}} \hat{f}(S)(n - 2|S|)\chi_S.$$

(c) Using the previous step, show that

$$\langle f, \overline{f} \rangle \le 4\mathbb{E}(Z)^2 + (n - 4)\mathbb{E}(Z^2).$$

(d) On the other hand, show (without using any of the above) that $\bar{f} \geq$ $(n-2)f$, from which it follows that

$$\langle f, \bar{f} \rangle \geq (n-2)\mathbb{E}(Z^2).$$

(e), Combine the last two steps to finish the argument.

6

First evidence of noise sensitivity of percolation

In this chapter, our goal is to collect some of the facts and theorems we have seen so far in order to conclude that percolation crossings are indeed noise sensitive. Recall from the "BKS" Theorem (Theorem 1.21) that it is enough for this purpose to prove that influences are "small" in the sense that $\sum_k \mathbf{I}_k(f_n)^2$ goes to 0. If we want to use only what we have actually proved in this book, namely Proposition 5.6, then we need to demonstrate (5.6) in this proposition.

In the first section, we deal with a careful study of influences in the case of percolation crossings on the triangular lattice. Then, we treat the case of $\mathbb{Z}^2$, where conformal invariance is not known. Finally, we speculate to what "extent" percolation is noise sensitive.

This whole chapter should be considered somewhat of a "pause" in our program, where we take the time to summarize what we have achieved so far in our understanding of the noise sensitivity of percolation, and what remains to be done if one wishes to obtain the exact "noise sensitivity exponent" as well as the existence of exceptional times for dynamical percolation.

6.1 Bounds on influences for crossing events in critical percolation on the triangular lattice

6.1.1 Setup

Fix $a, b > 0$, let us consider some rectangle $[0, a \cdot n] \times [0, b \cdot n]$, and let R_n be the set of hexagons in $\mathbb{T}$ that intersect $[0, a \cdot n] \times [0, b \cdot n]$. Let f_n be the event that there is a left-to-right crossing event in R_n. (This is the same event as in Example 1.22 in Chapter 1, but with $\mathbb{Z}^2$ replaced by $\mathbb{T}$.) By the RSW Theorem 2.1, we know that $\{f_n\}$ is nondegenerate. Conformal invariance tells us that $\mathbb{E}[f_n] = \mathbb{P}[f_n = 1]$ converges as $n \to \infty$. This limit is given by the so-called **Cardy's formula**.

To prove that this sequence of Boolean functions $\{f_n\}$ is noise sensitive, we wish to study its influence vector $\mathbf{Inf}(f_n)$ and we would like to prove that $\mathbf{H}(f_n) = \|\mathbf{Inf}(f_n)\|_2^2 = \sum \mathbf{I}_k(f_n)^2$ decays polynomially fast toward 0. (Recall that we have given a complete proof of Theorem 1.21 only in the case where $\mathbf{H}(f_n)$ decreases as an inverse polynomial of the number of variables.)

6.1.2 Study of the set of influences

Let x be a site (i.e., a hexagon) in the rectangle R_n. One needs to understand

$$\mathbf{I}_x(f_n) := \mathbb{P}[x \text{ is pivotal for } f_n]$$

It is easy but crucial to note that if x is at distance d from the boundary of R_n, in order for x to be pivotal, the *four-arm* event described in Chapter 2 (see Figure 2.4) has to be satisfied in the ball $B(x,d)$ of radius d around the hexagon x. See Figure 6.1.

In particular, this implies (still under the assumption that $\text{dist}(x, \partial R_n) = d$) that

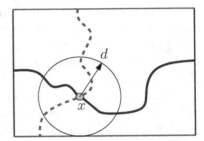

Figure 6.1 Illustration of the four-arm event.

$$\mathbf{I}_x(f_n) \le \alpha_4(d) = d^{-\frac{5}{4}+o(1)},$$

where $\alpha_4(d)$ denotes the probability of the four-arm event up to distance d. See Chapter 2. The statement

$$\alpha_4(R) = R^{-5/4+o(1)}$$

implies that for any $\epsilon > 0$, there exists a constant $C = C_\epsilon$, such that for all $R \ge 1$,

$$\alpha_4(R) \le C R^{-5/4+\epsilon}.$$

The above bound gives us very good control on the influences of the points in the *bulk* of the domain (i.e., the points far from the boundary). Indeed, for any fixed $\delta > 0$, let Δ_n^δ be the set of hexagons in R_n which are at distance at least δn from ∂R_n. Most of the points in R_n (except a proportion $O(\delta)$ of these) lie in Δ_n^δ, and for any such point $x \in \Delta_n^\delta$, one has by the above argument

$$\mathbf{I}_x(f_n) \le \alpha_4(\delta n) \le C(\delta n)^{-5/4+\epsilon} \le C\delta^{-5/4} n^{-5/4+\epsilon}. \tag{6.1}$$

Therefore, the contribution of these points to $\mathbf{H}(f_n) = \sum_k \mathbf{I}_k(f_n)^2$ is bounded by $O(n^2)(C\delta^{-5/4}n^{-5/4+\epsilon})^2 = O(\delta^{-5/2}n^{-1/2+2\epsilon})$. As $n \to \infty$, this goes to zero polynomially fast. Because this estimate concerns "almost" all points in R_n, it seems we are close to proving the strong form (5.6) of the BKS criterion. Still, to complete the above analysis, one has to estimate what the influences of the points near the boundary are.

6.1.3 Influence of the boundary

The main difficulty here is that if x is close to the boundary, the probability for x to be pivotal is no longer related to the above *four-arm* event. Think of the above figure when d gets very small compared to n. One has to distinguish two cases:

- x is close to a *corner*. This will correspond to a *two-arm* event in a quarter-plane.
- x is close to an *edge*. This involves the *three-arm* event in the half-plane $\mathbb{H}$.

Before detailing how to estimate the influence of points near the boundary, let us start by giving the necessary background on the involved critical exponents.

The two-arm and three-arm events in $\mathbb{H}$

For these particular events, it turns out that the critical exponents are known to be *universal*: they are two of the very few critical exponents that are known also on the square lattice $\mathbb{Z}^2$. The derivations of these exponents do not rely on SLE technology but are "elementary." Therefore, in this discussion, we consider both lattices $\mathbb{T}$ and $\mathbb{Z}^2$.

The *three-arm* event in $\mathbb{H}$ corresponds to the event that there are three arms (two open arms and one "closed" arm in the dual) going from 0 to distance R and such that they remain in the upper half-plane. See Figure 6.2 for a self-explanatory definition. The *two-arm* event corresponds to just having one open and one closed arm.

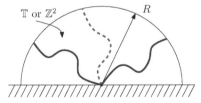

Figure 6.2 Illustration of the half-plane three-arm event.

Let $\alpha_2^+(R)$ and $\alpha_3^+(R)$ denote the probabilities of these events. As in Chapter 2, let $\alpha_2^+(r,R)$ and $\alpha_3^+(r,R)$ be the natural extensions to the annulus case (i.e., the probability that these events are satisfied in the annulus between radii r and R in the upper half-plane).

We will rely on the following result, which goes back as far as we know to M. Aizenman. See (Wer07) for a proof of this result.

Proposition 6.1 *Both on the triangular lattice $\mathbb{T}$ and on $\mathbb{Z}^2$, one has that*

$$\alpha_2^+(r,R) \asymp (r/R)$$

and

$$\alpha_3^+(r,R) \asymp (r/R)^2 .$$

Note that, in these special cases, there are no $o(1)$ correction terms in the exponent but rather these probabilities are known up to constants.

The two-arm event in the quarter-plane

In this case, the corresponding exponent is unfortunately not known on $\mathbb{Z}^2$, so we will need to do some work here in the next section, where we will prove noise sensitivity of percolation crossings on $\mathbb{Z}^2$.

The *two-arm* event in a corner corresponds to the event illustrated in Figure 6.3.

We will use the following proposition.

Proposition 6.2 (SW01) *If $\alpha_2^{++}(R)$ denotes the probability of this event, then*

$$\alpha_2^{++}(R) = R^{-2+o(1)} ,$$

and with the obvious notations

$$\alpha_2^{++}(r,R) = (r/R)^{2+o(1)} .$$

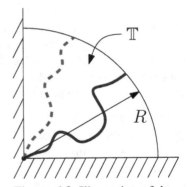

Figure 6.3 Illustration of the quarter-plane two-arm event.

Now, back to our study of influences, we are in good shape (at least for the triangular lattice) because the two critical exponents arising from the boundary effects are larger than the *bulk* exponent $5/4$. This means that it is less likely for a point near the boundary to be pivotal than for a point in the bulk. Therefore in some sense the boundary helps us here.

More formally, summarizing the above facts, for any $\epsilon > 0$, there is a constant $C = C(\epsilon)$ such that for any $1 \leq r \leq R$,

$$\max\{\alpha_4(r,R), \alpha_3^+(r,R), \alpha_2^{++}(r,R)\} \leq C(r/R)^{\frac{5}{4}-\epsilon}. \tag{6.2}$$

Now, if x is some hexagon in R_n, let n_0 be the distance to the closest edge of ∂R_n and let x_0 be the point on ∂R_n such that $\text{dist}(x, x_0) = n_0$. Next, let $n_1 \geq n_0$ be the distance from x_0 to the closest corner and let x_1 be this closest corner. It is easy to see that for x to be pivotal for f_n, the following events all have to be satisfied:

- The four-arm event in the ball of radius n_0 around x
- The $\mathbb{H}$-three-arm event in the annulus centered at x_0 of radii $2n_0$ and n_1
- The corner-two-arm event in the annulus centered at x_1 of radii $2n_1$ and n

By independence on disjoint sets, one thus concludes that

$$\mathbf{I}_x(f_n) \leq \alpha_4(n_0)\alpha_3^+(2n_0, n_1)\alpha_2^{++}(2n_1, n)$$
$$\leq O(1)n^{-5/4+\epsilon}.$$

6.1.4 Noise sensitivity of crossing events

The uniform bound (6.3) on the influences over the whole domain R_n enables us to conclude that the strong form (5.6) of the BKS criterion is satisfied. Indeed,

$$\mathbf{H}(f_n) = \sum_{x \in R_n} \mathbf{I}_x(f_n)^2 \leq Cn^2(n^{-5/4+\epsilon})^2 = Cn^{-1/2+2\epsilon}, \tag{6.3}$$

where $C = C(a, b, \epsilon)$ is a universal constant. By taking $\epsilon < 1/4$, this gives us the desired polynomial decay on $\mathbf{H}(f_n)$, which by Proposition 5.6 implies

Theorem 6.3 (BKS99) *The sequence of percolation crossing events $\{f_n\}$ on $\mathbb{T}$ is noise sensitive.*

We give some other consequences (e.g., to sharp thresholds) of the preceding analysis on the influences of the crossing events in a later section in this chapter.

6.2 The case of $\mathbb{Z}^2$ percolation

Let R_n denote similarly the $\mathbb{Z}^2$ rectangle closest to $[0, a \cdot n] \times [0, b \cdot n]$ and let f_n be the corresponding left–right crossing event (so here this corresponds exactly to Example 1.22). Here one has to face two main difficulties:

- The main one is that due to the missing ingredient of *conformal invariance*, we do not have at our disposal the value of the *four-arm* critical exponent (which is of course believed to be $5/4$). In fact, even the *existence* of a critical exponent is an open problem.
- The second difficulty (also due to the lack of conformal invariance) is that it is now slightly harder to deal with boundary issues. Indeed, we can still use the above bounds on α_3^+ which are *universal*, but the exponent 2 for α_2^{++} is not known for $\mathbb{Z}^2$. So this requires some more analysis.

Let us start by taking care of the boundary effects.

6.2.1 Handling the boundary effect

What we need to do to carry through the above analysis for $\mathbb{Z}^2$ is to obtain a reasonable estimate on α_2^{++}. Fortunately, the following bound, which follows immediately from Proposition 6.1, is sufficient:

$$\alpha_2^{++}(r, R) \le O(1) \frac{r}{R}. \tag{6.4}$$

Now let e be an edge in R_n. We wish to bound from above $\mathbf{I}_e(f_n)$. We use the same notation as in the case of the triangular lattice: recall the definitions of n_0, x_0, n_1, x_1 there.

We obtain in the same way

$$\mathbf{I}_e(f_n) \le \alpha_4(n_0) \alpha_3^+(2n_0, n_1) \alpha_2^{++}(2n_1, n). \tag{6.5}$$

At this point, we need another *universal* exponent, which goes back also to M. Aizenman:

Theorem 6.4 (M. Aizenman, see (Wer07)) *Let $\alpha_5(r, R)$ denote the probability that there are five arms (with four of them being of "alternate colors"). Then there are universal constants $c, C > 0$ such that both for $\mathbb{T}$ and $\mathbb{Z}^2$, one has for all $1 \le r \le R$,*

$$c(\frac{r}{R})^2 \le \alpha_5(r, R) \le C(\frac{r}{R})^2.$$

This result allows us to get a lower bound on $\alpha_4(r,R)$. Indeed, it is clear that

$$\alpha_4(r,R) \geq \alpha_5(r,R) \geq \Omega(1)\alpha_3^+(r,R). \qquad (6.6)$$

In fact, one can obtain the following better lower bound on $\alpha_4(r,R)$ which we will need later.

Lemma 6.5 *There exists some $\epsilon > 0$ and some constant $c > 0$ such that for any $1 \leq r \leq R$,*

$$\alpha_4(r,R) \geq c(r/R)^{2-\epsilon}.$$

Proof There are several ways to see why this holds, none of them being either very hard or very easy. One of them is to use **Reimer's inequality** (see (Re00; Gri99)) which in this case would imply that

$$\alpha_5(r,R) \leq \alpha_1(r,R)\alpha_4(r,R). \qquad (6.7)$$

The RSW Theorem 2.1 can be used to show that

$$\alpha_1(r,R) \leq (r/R)^{\alpha}$$

for some positive α. By Theorem 6.4, we are done. [See [(GPS10), Section 2.2 as well as the appendix] for more on these bounds.] $\square$

Combining (6.5) with (6.6), one obtains

$$\mathbf{I}_e(f_n) \leq O(1)\alpha_4(n_0)\alpha_4(2n_0,n_1)\alpha_2^{++}(2n_1,n)$$
$$\leq O(1)\alpha_4(n_1)\frac{n_1}{n},$$

where in the last inequality we used quasi-multiplicativity (Proposition 2.3) as well as the bound given by (6.4).

Recall that we want an upper bound on $\mathbf{H}(f_n) = \sum \mathbf{I}_e(f_n)^2$. In this sum over edges $e \in R_n$, let us divide the set of edges into dyadic annuli centered around the four corners as in the next picture.

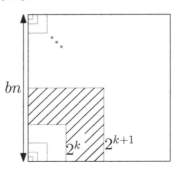

Notice that there are $O(1)2^{2k}$ edges in an annulus of radius 2^k. This enables us to bound $\mathbf{H}(f_n)$ as follows:

$$\sum_{e \in R_n} \mathbf{I}_e(f_n)^2 \leq O(1) \sum_{k=1}^{\log_2 n + O(1)} 2^{2k} \left(\alpha_4(2^k) \frac{2^k}{n} \right)^2$$

$$\leq O(1) \frac{1}{n^2} \sum_{k \leq \log_2 n + O(1)} 2^{4k} \alpha_4(2^k)^2. \tag{6.8}$$

It now remains to obtain a good upper bound on $\alpha_4(R)$, for all $R \geq 1$.

6.2.2 An upper bound on the four-arm event in $\mathbb{Z}^2$

This turns out to be a rather nontrivial problem. Recall that we obtained an easy lower bound on α_4 using α_5 (and Lemma 6.5 strengthens this lower bound). For an upper bound, completely different ideas are required. On $\mathbb{Z}^2$, the following estimate is available for the four-arm event.

Proposition 6.6   *For critical percolation on $\mathbb{Z}^2$, there exist constants $\epsilon, C > 0$ such that for any $R \geq 1$, one has*

$$\alpha_4(1, R) \leq C \left(\frac{1}{R} \right)^{1+\epsilon}.$$

Before discussing where such an estimate comes from, let us see that it indeed implies a polynomial decay for $\mathbf{H}(f_n)$.

Recalling Equation (6.8) and plugging in the above estimate, this gives us

$$\sum_{e \in R_n} \mathbf{I}_e(f_n)^2 \leq O(1) \frac{1}{n^2} \sum_{k \leq \log_2 n + O(1)} 2^{4k} (2^k)^{-2-2\epsilon}$$

$$\leq O(1) \frac{1}{n^2} n^{2-2\epsilon} = O(1) n^{-2\epsilon},$$

which implies the desired polynomial decay and thus the fact that $\{f_n\}$ is noise sensitive by Proposition 5.6.

Let us now discuss different approaches that allow one to prove Proposition 6.6.

(1) Kesten proved implicitly this estimate in his celebrated paper (Kes87). His main motivation for such an estimate was to obtain bounds on the corresponding critical exponent that governs the so-called *critical length*.

(2) In (BKS99), to prove noise sensitivity of percolation using their criterion on $\mathbf{H}(f_n)$, the authors referred to (Kes87), but they also gave a completely different approach that also yields this estimate.

   Their alternative approach is very nice: finding an upper bound for $\alpha_4(R)$ is related to finding an upper bound for the influences for crossings of an $R \times R$ box. For this, they noticed the following nice phenomenon: if a monotone function f happens to be very little correlated with Majority, then its influences have to be small (see Sections 12.12 and 12.13). The proof of this phenomenon uses for the first time in this context the concept of "randomized algorithms."

(3) In (SS10), the concept of randomized algorithms is used in a more powerful way; see Chapter 8. In this chapter, a proof of Proposition 6.6 is provided.

Remark It turns out that that a multiscale version of Proposition 6.6 stating that $\alpha_4(r, R) \leq C\left(\frac{r}{R}\right)^{1+\epsilon}$ is also true. However, none of the three arguments given above seem to prove this stronger version. A proof of this stronger version is given in the appendix of (SS11). Because this multiscale version is not needed until Chapter 10, we stated here only the weaker version.

6.3 Some other consequences of our study of influences

In the previous sections, we handled the boundary effects to check that $\mathbf{H}(f_n)$ indeed decays polynomially fast. Let us list some related results implied by this analysis.

6.3.1 Energy spectrum of f_n

We start by a straightforward observation: because the f_n are monotone, we have by Proposition 4.5 that

$$\widehat{f_n}(\{x\}) = \frac{1}{2}\mathbf{I}_x(f_n),$$

for any site x (or edge e) in R_n. Therefore, the bounds we obtained on $\mathbf{H}(f_n)$ imply the following control on the first layer of the energy spectrum of the crossing events $\{f_n\}$.

Corollary 6.7 *Let $\{f_n\}$ be the crossing events of the rectangles R_n.*

(i) If we are on the triangular lattice $\mathbb{T}$, then we have the bound

$$E_{f_n}(1) = \sum_{|S|=1} \widehat{f_n}(S)^2 \leq n^{-1/2+o(1)}.$$

(ii) On the square lattice $\mathbb{Z}^2$, we end up with the weaker estimate

$$E_{f_n}(1) \le C n^{-\epsilon},$$

for some $\epsilon, C > 0$.

6.3.2 Sharp threshold of percolation

The preceding analysis gave an upper bound on $\sum_k I_k(f_n)^2$. As we have seen in the first chapters, the total influence $I(f_n) = \sum_k I_k(f_n)$ is also a very interesting quantity. Recall that, by Russo's formula, this is the quantity that shows "how sharp" the threshold is for $p \mapsto \mathbb{P}_p[f_n = 1]$.

The preceding analysis allows us to prove the following.

Proposition 6.8   *Both on $\mathbb{T}$ and $\mathbb{Z}^2$, one has*

$$I(f_n) \asymp n^2 \alpha_4(n).$$

In particular, this shows that on $\mathbb{T}$ that

$$I(f_n) \asymp n^{3/4 + o(1)}.$$

Remark Since f_n is defined on $\{-1, 1\}^{O(n^2)}$, note that the Majority function defined on the same hypercube has a much sharper threshold than the percolation crossings f_n.

Proof   We first derive an upper bound on the total influence. In the same vein (i.e., using dyadic annuli and quasi-multiplicativity) as we derived (6.8) and with the same notation one has

$$I(f_n) = \sum_e I_e(f_n) \le \sum_e O(1)\alpha_4(n_1)\frac{n_1}{n}$$

$$\le O(1)\frac{1}{n} \sum_{k \le \log_2 n + O(1)} 2^{3k}\alpha_4(2^k).$$

Now, and this is the main step here, using quasi-multiplicativity one has $\alpha_4(2^k) \le O(1)\frac{\alpha_4(n)}{\alpha_4(2^k, n)}$, which gives us

$$\mathbf{I}(f_n) \le O(1)\frac{\alpha_4(n)}{n} \sum_{k \le \log_2 n + O(1)} 2^{3k} \frac{1}{\alpha_4(2^k, n)}$$

$$\le O(1)\frac{\alpha_4(n)}{n} \sum_{k \le \log_2 n + O(1)} 2^{3k} \frac{n^2}{2^{2k}} \quad \text{since } \alpha_4(r, R) \ge \alpha_5(r, R) \asymp (r/R)^{-2}$$

$$\le O(1)n\alpha_4(n) \sum_{k \le \log_2 n + O(1)} 2^k$$

$$\le O(1)n^2\alpha_4(n)$$

as desired.

For the lower bound on the total influence, we proceed as follows. One obtains a lower bound by just summing over the influences of points whose distance to the boundary is at least $n/4$. It would suffice if we knew that for such edges or hexagons, the influence is at least a constant times $\alpha_4(n)$. This is in fact known to be true. It is not very involved and is part of the *folklore* results in percolation. However, it still would lead us too far from our topic. The needed technique is known under the name of **separation of arms** and is clearly related to the statement of quasi-multiplicativity. See (Wer07) for more details. □

6.4 Quantitative noise sensitivity

In this chapter, we have proved that the sequence of crossing events $\{f_n\}$ is noise sensitive. This can be roughly translated as follows: for any fixed level of noise $\epsilon > 0$, as $n \to \infty$, the large-scale clusters of ω in the window $[0, n]^2$ are asymptotically independent of the large-scale clusters of ω_ϵ.

Remark Note that this picture is correct, but to make it rigorous would require some work, because so far we worked only with left–right crossing events. The nontrivial step here is to prove that in some sense, in the scaling limit $n \to \infty$, any macroscopic property concerning percolation (e.g., diameter of clusters) is measurable with respect to the σ-algebra generated by the crossing events. This is a rather subtle problem because we need to make precise what kind of information we keep in what we call the "scaling limit" of percolation (or subsequential scaling limits in the case of $\mathbb{Z}^2$). An example of something that is not present in the scaling limit is whether one has more open sites than closed ones because by noise sensitivity we know that this is asymptotically uncorrelated with crossing events. We will not need to discuss these notions of scaling limits more here because the focus is mainly on the discrete model itself, including the model of dynamical percolation which is presented in Chapter 11.

At this stage, a natural question to ask is to what extent the percolation picture is sensitive to "small" noise. In other words, can we let the noise $\epsilon = \epsilon_n$ go to zero with the "size of the system" n, and yet keep this independence of large-scale structures between ω and ω_{ϵ_n}? If yes, can we give *quantitative estimates* on how fast the noise $\epsilon = \epsilon_n$ may go to zero? One can state this question more precisely as follows.

Question 6.9 *If $\{f_n\}$ corresponds to our left–right crossing events, for which sequences of noise levels $\{\epsilon_n\}$ do we have*

$$\lim_{n\to\infty} \mathrm{Cov}[f_n(\omega), f_n(\omega_{\epsilon_n})] = 0 \, ?$$

The purpose of this section is to briefly discuss this question based on the results we have obtained so far.

6.4.1 Link with the energy spectrum of $\{f_n\}$

It is an exercise to show that Question 6.9 is essentially equivalent to the following one (see Exercise 9.1 in Chapter 9).

Question 6.10 *For which sequences $\{k_n\}$ going to infinity do we have*

$$\sum_{m=1}^{k_n} E_{f_n}(m) = \sum_{1\le|S|\le k_n} \widehat{f_n}(S)^2 \xrightarrow[n\to\infty]{} 0 \, ?$$

Recall that we have already obtained some relevant information on this question. Indeed, we have proved in this chapter that $\mathbf{H}(f_n) = \sum_x \mathbf{I}_x(f_n)^2$ decays polynomially fast toward 0 (both on $\mathbb{Z}^2$ and $\mathbb{T}$). Therefore Proposition 5.6 tells us that for some constant $c > 0$, one has for both $\mathbb{T}$ and $\mathbb{Z}^2$ that

$$\sum_{1\le|S|\le c\log n} \widehat{f_n}(S)^2 \to 0. \tag{6.9}$$

Therefore, back to our original question 6.9, this gives us the following quantitative statement: if the noise ϵ_n satisfies $\epsilon_n \gg \frac{1}{\log n}$, then $f_n(\omega)$ and $f_n(\omega_{\epsilon_n})$ are asymptotically independent.

6.4.2 Noise stability regime

Of course, one cannot be too demanding on the rate of decay of $\{\epsilon_n\}$. For example if $\epsilon_n \ll \frac{1}{n^2}$, then in the domain $[0,n]^2$, with high probability, the configurations ω and ω_{ϵ_n} are identical. This brings us to the next natural question concerning the *noise stability regime* of crossing events.

Question 6.11 *Let $\{f_n\}$ be our sequence of crossing events. For which sequences $\{\epsilon_n\}$ do we have*

$$\mathbb{P}[f_n(\omega) \neq f_n(\omega_{\epsilon_n})] \underset{n \to \infty}{\longrightarrow} 0 \; ?$$

It is again an exercise to show that this question is essentially equivalent to the following one (see Exercise 9.2 in Chapter 9).

For which sequences $\{k_n\}$ do we have

$$\sum_{|S| > k_n} \widehat{f_n}(S)^2 \to 0 \; ?$$

Using the estimates of the present chapter, one can give the following nontrivial bound on the noise stability regime of $\{f_n\}$.

Proposition 6.12 *Both on $\mathbb{Z}^2$ and $\mathbb{T}$, if*

$$\epsilon_n = o\left(\frac{1}{n^2 \alpha_4(n)}\right),$$

then

$$\mathbb{P}[f_n(\omega) \neq f_n(\omega_{\epsilon_n})] \underset{n \to \infty}{\longrightarrow} 0$$

On the triangular grid, using the critical exponent, this gives us a bound of $n^{-3/4}$ on the noise stability regime of percolation.

Proof Let $\{\epsilon_n\}$ be a sequence satisfying the above assumption. There are $O(n^2)$ bits concerned. For simplicity, assume that there are exactly n^2 bits.

Let $\omega = \omega_0 = (x_1, \dots, x_{n^2})$ be sampled according to the uniform measure. Recall that the noised configuration ω_{ϵ_n} is produced as follows: for each $i \in [n^2]$, resample the value x_i with probability ϵ_n, independently of everything else, obtaining the value y_i. (In particular $y_i \neq x_i$ with probability $\epsilon_n/2$).

Now for each $i \in [n^2]$ define the intermediate configuration

$$\omega_i := (y_1, \dots, y_i, x_{i+1}, \dots, x_{n^2})$$

Notice that for each $i \in [n^2]$, ω_i is also sampled according to the uniform measure and one has for each $i \in \{1, \dots, n^2\}$ that

$$\mathbb{P}[f_n(\omega_{i-1}) \neq f_n(\omega_i)] = (\epsilon_n/2) \mathbf{I}_i(f_n).$$

Summing over all i, one obtains

$$\mathbb{P}[f_n(\omega) \neq f_n(\omega_{\epsilon_n})] = \mathbb{P}[f_n(\omega_0) \neq f_n(\omega_{n^2})]$$

$$\leq \sum_{i=1}^{n^2} \mathbb{P}[f_n(\omega_{i-1}) \neq f_n(\omega_i)]$$

$$= (\epsilon_n/2) \sum_{i=1}^{n^2} \mathbf{I}_i(f_n)$$

$$= (\epsilon_n/2) \mathbf{I}(f_n)$$

$$\leq \epsilon_n O(1) n^2 \alpha_4(n) \quad \text{by Proposition 6.8,}$$

which concludes the proof. □

Remark The proof of Proposition 6.12 immediately yields that given any sequence of Boolean functions $\{f_n\}$, if $\epsilon_n \mathbf{I}(f_n) \to 0$ as $n \to \infty$, then $\mathbb{P}[f_n(\omega) \neq f_n(\omega_{\epsilon_n})] \to 0$ as $n \to \infty$.

It might be of interest to see a different proof of Proposition 6.12 that also holds in the context of the previous remark.

Alternative proof of Proposition 6.12

We assume the functions $\{f_n\}$ map into $\{\pm 1\}$. We then have

$$\mathbb{E}[f_n(\omega) f_n(\omega_{\epsilon_n})] = \sum_S \hat{f}(S)^2 (1 - \epsilon_n)^{|S|}$$

$$\geq (1 - \epsilon_n)^{\sum_S |S| \hat{f}(S)^2}$$

$$= (1 - \epsilon_n)^{\mathbf{I}(f_n)}$$

$$(6.10)$$

where the inequality follows from Jensen's inequality and the last equality from Proposition 4.4 (and the remark afterwards). The assumption yields that the last term approaches 1 as $n \to \infty$. This is equivalent to the statement of the result. □

6.4.3 *Where does the spectral mass lie?*

Proposition 6.12 (together with Exercise 9.2 in Chapter 9) implies that the Fourier coefficients of $\{f_n\}$ satisfy

$$\sum_{|S| \gg n^2 \alpha_4(n)} \hat{f_n}(S)^2 \xrightarrow[n \to \infty]{} 0. \qquad (6.11)$$

From Lemma 6.5, we know that even on $\mathbb{Z}^2$, $n^2 \alpha_4(n)$ is larger than n^ϵ for some exponent $\epsilon > 0$. Combining the estimates on the spectrum that we achieved so far (equations (6.9) and (6.11)), we see that to localize the spectral mass of $\{f_n\}$, there is still a missing gap. See Figure 6.4.

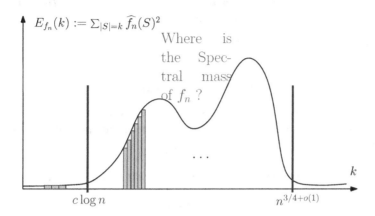

Figure 6.4 This picture summarizes our present knowledge of the energy spectrum of $\{f_n\}$ on the triangular lattice $\mathbb{T}$. Much remains to be understood to know where, in the range $[\Omega(\log n), n^{3/4+o(1)}]$, the spectral mass lies. This question is analyzed in the following chapters.

For our later applications to the model of dynamical percolation in Chapter 11, an understanding of the noise sensitivity of percolation better than the "logarithmic" control we have achieved so far will be needed.

6.5 Exercises

Instead of the usual set of exercises, this chapter ends with a single problem whose goal is to do hands-on computations of the first layers of the energy spectrum of the percolation crossing events f_n. Recall from Proposition 4.2 that a sequence of Boolean functions $\{f_n\}$ is noise sensitive if and only if for any fixed $k \geq 1$,

$$\sum_{m=1}^{k} \sum_{|S|=m} \widehat{f_n}(S)^2 = \sum_{m=1}^{k} E_{f_n}(m) \xrightarrow[n\to\infty]{} 0.$$

In this chapter, we verified (using Proposition 4.5) that this is indeed the case for $k = 1$. The purpose here is to check by simple combinatorial arguments (without relying on hypercontractivity) that it is still the case for $k = 2$ and to convince ourselves that it works for all layers $k \geq 3$.

To start with, we simplify our task by working on the torus $\mathbb{Z}^2/n\mathbb{Z}^2$. This has the great benefit of eliminating boundary issues.

Energy spectrum of crossing events on the torus (study of the first layers)

Let T_n be either the square grid torus $\mathbb{Z}^2/n\mathbb{Z}^2$ or the triangular grid torus $\mathbb{T}/n\mathbb{T}$. Let f_n be the indicator of the event that there is an open circuit along the first coordinate of T_n.

(a) Using RSW, prove that there is a constant $c > 0$ such that for all $n \geq 1$,

$$c \leq \mathbb{P}[f_n = 1] \leq 1 - c.$$

(In other words, $\{f_n\}$ is nondegenerate.)

(b) Show that for all edges e (or sites x) in T_n

$$\mathbf{I}_e(f_n) \leq \alpha_4(\frac{n}{2}).$$

(c) Check that the BKS criterion (about $\mathbf{H}(f_n)$) is satisfied. Therefore $\{f_n\}$ is noise sensitive.

From now on, one would like to forget about the BKS Theorem and try to do some hands-on computations to get a feeling why most frequencies should be large.

(d) Show that if x, y are two sites of T_n (or similarly if e, e' are two edges of T_n), then

$$|\hat{f}(\{x,y\})| \leq 2\mathbb{P}[\, x \text{ and } y \text{ are pivotal points}].$$

Does this result hold for general Boolean functions?

(e) Show that if $d := |x - y|$, then

$$\mathbb{P}[\, x \text{ and } y \text{ are pivotal points}\,] \leq O(1)\frac{\alpha_4(n/2)^2}{\alpha_4(\frac{d}{2},\frac{n}{2})}.$$

(*Hint:* Use Proposition 2.3.)

(f) On the square lattice $\mathbb{Z}^2$, by carefully summing over all edges $e, e' \in T_n \times T_n$, show that

$$E_{f_n}(2) = \sum_{|S|=2} \widehat{f_n}(S)^2 \le O(1)n^{-\epsilon},$$

for some exponent $\epsilon > 0$.

Hint: You might decompose the sum in a dyadic way (as we did many times in the present section) depending on the mutual distance $d(e, e')$.

(g) On the triangular grid, what exponent does it give for the decay of $\mathbb{E}_{f_n}(2)$? Compare with the decay we found in Corollary 6.7 about the decay of the first layer $E_{f_n}(1)$ (i.e. $k = 1$). See also Lemma 5.7 in this regard. Discuss this.

(h) For $\mathbb{T}$, what do you expect for higher (fixed) values of k? (i.e., for $E_{f_n}(k), k \ge 3$)?

Observe that one can do similar things for rectangles but then one has to deal with boundary issues.

7

Anomalous fluctuations

In this chapter, our goal is to extend the technology we used to prove the KKL Theorems on influences and the BKS Theorem on noise sensitivity in a slightly different context: the study of fluctuations in **first-passage percolation**.

7.1 The model of first-passage percolation

Let us first explain what the model is. Let $0 < a < b$ be two positive numbers. We define a **random metric** on the graph $\mathbb{Z}^d$, $d \geq 2$ as follows. Independently for each edge $e \in \mathbb{E}^d$, fix its length τ_e to be a with probability $1/2$ and b with probability $1/2$. This is represented by a uniform configuration $\omega \in \{-1, 1\}^{\mathbb{E}^d}$.

This procedure induces a well-defined (random) metric dist_ω on $\mathbb{Z}^d$ in the usual fashion. For any vertices $x, y \in \mathbb{Z}^d$, let

$$\text{dist}_\omega(x, y) := \inf_{\substack{\text{paths } \gamma = \{e_1, \ldots, e_k\} \\ \text{connecting } x \to y}} \left\{ \sum \tau_{e_i}(\omega) \right\}.$$

Remark In greater generality, the lengths of the edges are i.i.d. nonnegative random variables, but here, following (BKS03), we restrict ourselves to the above uniform distribution on $\{a, b\}$ to simplify the exposition; see (BR08) for an extension to more general laws.

One of the main goals in first-passage percolation is to understand the large-scale properties of this random metric space. For example, for any $T \geq 1$, one may consider the (random) ball

$$B_\omega(x, T) := \{y \in \mathbb{Z}^d : \text{dist}_\omega(x, y) \leq T\}.$$

To understand the name *first-passage percolation*, one can think of this model as follows. Imagine that water is pumped in at vertex x, and that for

74

each edge e, it takes $\tau_e(\omega)$ units of time for the water to travel across the edge e. Then, $B_\omega(x, T)$ represents the region of space that has been wetted by time T.

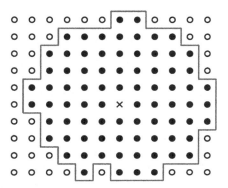

Figure 7.1 A sample of a wetted region at time T, i.e., $B_\omega(x, T)$, in first-passage percolation.

An application of subadditivity shows that the renormalized ball $\frac{1}{T} B_\omega(0, T)$ converges as $T \to \infty$ toward a deterministic shape that can in certain cases be computed explicitly. This is a kind of "geometric law of large numbers". Whence the natural question:

Question 7.1 *Describe the* fluctuations *of* $B_\omega(0, T)$ *around its asymptotic deterministic shape.*

This question has received tremendous interest in the last 15 years or so. It is widely believed that these fluctuations should be in some sense "universal." More precisely, the behavior of $B_\omega(0, T)$ around its limiting shape should not depend on the "microscopic" particularities of the model such as the law on the edges lengths but only on the dimension d of the underlying graph. The shape itself depends on the other hand of course on the microscopic parameters, in the same way as the critical point depends on the graph in percolation.

In the two-dimensional case, using very beautiful combinatorial bijections with random matrices, certain cases of *directed* last passage percolation (where the law on the edges is taken to be geometric or exponential) have been understood very deeply. For example, it is known (see (Joh00)) that the fluctuations of the ball of radius n (i.e., the points whose last passage times are below n) around n times its asymptotic deterministic shape are of order $n^{1/3}$ and the law of these fluctuations properly renormalized follows the Tracy–Widom distribution. Very interestingly, the fluctuations of the largest eigenvalue of GUE ensembles also follow this distribution.

7.2 State of the art

Returning to our initial model of (nondirected) first passage percolation, it is thus conjectured that, for dimension $d = 2$, fluctuations are of order $n^{1/3}$ following the Tracy–Widom law. Still, the current state of understanding of this model is far from this conjecture.

Kesten first proved that the fluctuations of the ball of radius n are at most $\sqrt{n}$ (this did not yet exclude a possible Gaussian behavior with Gaussian scaling). Benjamini, Kalai, and Schramm then strengthened this result by showing that the fluctuations are sub-Gaussian. This is still far from the conjectured $n^{1/3}$-fluctuations, but their approach has the great advantage of being very general; in particular their result holds in any dimension $d \geq 2$.

Let us now state their main theorem concerning the fluctuations of the metric dist.

Theorem 7.2 (BKS03) *For all a,b,d, there exists an absolute constant $C = C(a,b,d)$ such that in $\mathbb{Z}^d$,*

$$\text{Var}(\text{dist}_\omega(0,v)) \leq C \frac{|v|}{\log|v|}$$

for any $v \in \mathbb{Z}^d, |v| \geq 2$.

To keep things simple here, we prove the analogous statement on the torus only where one has more symmetries and invariance to play with.

7.3 The case of the torus

Let $\mathbb{T}_m^d$ be the d-dimensional torus $(\mathbb{Z}/m\mathbb{Z})^d$. As in the preceding lattice model, independently for each edge of $\mathbb{T}_m^d$, we choose its length to be either a or b equally likely. We are interested here in the smallest length among all closed paths γ "winding" around the torus along the first coordinate $\mathbb{Z}/m\mathbb{Z}$ (i.e., those paths γ that when projected onto the first coordinate have winding number one). In (BKS03), this is called the shortest *circumference*. For any configuration $\omega \in \{a,b\}^{E(\mathbb{T}_m^d)}$, this shortest circumference is denoted by $\text{Circ}_m(\omega)$.

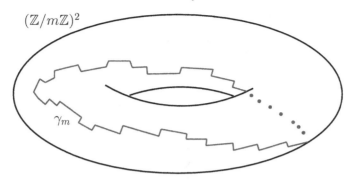

$(\mathbb{Z}/m\mathbb{Z})^2$

γ_m

Figure 7.2 The shortest geodesic along the first coordinate for the random metric dist_ω on $(\mathbb{Z}/m\mathbb{Z})^2$.

Theorem 7.3 (BKS03) *There is a constant $C = C(a,b)$ (which does not depend on the dimension d), such that*

$$\mathrm{var}(\mathrm{Circ}_m(\omega)) \leq C\frac{m}{\log m}.$$

Remark A similar analysis as the one carried out below works in greater generality: if $G = (V,E)$ is some finite connected graph endowed with a random metric d_ω with $\omega \in \{a,b\}^{\otimes E}$, then one can obtain bounds on the fluctuations of the random diameter $D = D_\omega$ of (G,d_ω). See (BKS03, Theorem 2) for a precise statement in this more general context.

Proof For any edge e, let us consider the gradient along the edge e: $\nabla_e \mathrm{Circ}_m$. These gradient functions have values in $[-(b-a),b-a]$. By dividing our distances by the constant factor $b-a$, we can even assume without loss of generality that our gradient functions have values in $[-1,1]$. Doing so, we end up being in a setup similar to the one we had in Chapter 5. The **influence** of an edge e corresponds here to $\mathbf{I}_e(\mathrm{Circ}_m) := \mathbb{P}[\nabla_e\mathrm{Circ}_m(\omega) \neq 0]$. We prove later on that Circ_m has very small influences. In other words, we will show that the above gradient functions have small support, and hypercontractivity will imply the desired bound.

We have thus reduced the problem to the following general framework. Consider a real-valued function $f: \{-1,1\}^n \to \mathbb{R}$, such that for any variable k, $\nabla_k f \in [-1,1]$. We are interested in $\mathrm{Var}(f)$ and we want to show that if "influences are small" then $\mathrm{Var}(f)$ is small. It is easy to check that the variance can be written

$$\mathrm{Var}(f) = \frac{1}{4} \sum_{k} \sum_{\emptyset \neq S \subseteq [n]} \frac{1}{|S|} \widehat{\nabla_k f}(S)^2.$$

If all the variables have very small influence, then, as previously, $\nabla_k f$ should be of high frequency. Heuristically, this should then imply that

$$\mathrm{Var}(f) \ll \sum_{k} \sum_{S \neq \emptyset} \widehat{\nabla_k f}(S)^2$$

$$= \sum_{k} \mathbf{I}_k(f).$$

This intuition is quantified by the following lemma on the link between the fluctuations of a real-valued function f on Ω_n and its influence vector.

Lemma 7.4 *Let $f : \Omega_n \to \mathbb{R}$ be a (real-valued) function such that each of its discrete derivatives $\nabla_k f$, $k \in [n]$ have their values in $[-1, 1]$. Let $\mathbf{I}_k(f) := \mathbb{P}[\nabla_k f \neq 0]$ be the influence of the kth bit. Assume that the influences of f are small in the sense that there exists some $\alpha > 0$ such that for any $k \in \{1, \ldots, n\}$, $\mathbf{I}_k(f) \leq n^{-\alpha}$. Then there is some constant $C = C(\alpha)$ such that*

$$\mathrm{Var}(f) \leq \frac{C}{\log n} \sum_{k} \mathbf{I}_k(f).$$

Remark If f is Boolean, then this follows from Theorem 1.16 with $C(\alpha) = c/\alpha$ with c universal.

The proof of this lemma is postponed to the next section. In the meantime, let us show that in our special case of first-passage percolation on the torus, the assumption on small influences is indeed verified. Because the edge lengths are in $\{a, b\}$, the smallest contour $\mathrm{Circ}_m(\omega)$ in $\mathbb{T}_m^d$ around the first coordinate lies somewhere in $[am, bm]$. Hence, if γ is a geodesic (a path in the torus with the required winding number) satisfying $\mathrm{length}(\gamma) = \mathrm{Circ}_m(\omega)$, then γ uses at most $\frac{b}{a}m$ edges. There might be several different geodesics minimizing the circumference. Let us choose randomly one of these in an "invariant" way and call it $\tilde{\gamma}$. For any edge $e \in E(\mathbb{T}_m^d)$, if, by changing the length of e, the circumference increases, then e has to be contained in any geodesic γ, and in particular in $\tilde{\gamma}$. This implies that $\mathbb{P}[\nabla_e \mathrm{Circ}_m(\omega) < 0] \leq \mathbb{P}[e \in \tilde{\gamma}]$. By symmetry we obtain

$$\mathbf{I}_e(\mathrm{Circ}_m) = \mathbb{P}[\nabla_e \mathrm{Circ}_m(\omega) \neq 0] \leq 2\mathbb{P}[e \in \tilde{\gamma}].$$

Now using the symmetries both of the torus $\mathbb{T}_m^d$ and of our observable Circ_m, if $\tilde{\gamma}$ is chosen in an appropriate invariant way (uniformly among all geodesics, for instance), then it is clear that all the "vertical" edges (meaning those edges that, when projected onto the first coordinate, project onto a single vertex) have the same probability to lie in $\tilde{\gamma}$. The same is true for the "horizontal" edges. In particular we have that

$$\sum_{\substack{\text{"vertical" edges } e}} \mathbb{P}[e \in \tilde{\gamma}] \le \mathbb{E}[|\tilde{\gamma}|] \le \frac{b}{a}m.$$

Because there are at least order m^d vertical edges, the influence of each of these is bounded by $O(1)m^{1-d}$. The same is true for the horizontal edges. All together this gives the desired assumption needed in Lemma 7.4. Applying this lemma, we indeed obtain

$$\text{Var}(\text{Circ}_m(\omega)) \le O(1)\frac{m}{\log m},$$

where the constant does not depend on the dimension d; the dimension in fact helps us here, because it makes the influences smaller. □

Remark At this point, we know that for any edge e, $\mathbf{I}_e(\text{Circ}_m) = O(\frac{m}{m^d})$. Hence, at least in the case of the torus, one easily deduces from Poincaré's inequality the theorem by Kesten, which says that $\text{Var}(\text{Circ}_m) = O(m)$.

7.4 Upper bounds on fluctuations in the spirit of KKL

In this section, we prove Lemma 7.4.

Proof Similarly as in the proofs of Chapter 5, the proof relies on implementing hypercontractivity in the right way. We have that for any c,

$$\text{var}(f) = \frac{1}{4}\sum_k \sum_{S \neq \emptyset} \frac{1}{|S|}\widehat{\nabla_k f}(S)^2$$

$$\le \frac{1}{4}\sum_k \sum_{0<|S|<c\log n} \widehat{\nabla_k f}(S)^2 + \frac{O(1)}{\log n}\sum_k \mathbf{I}_k(f)$$

where the $O(1)$ term depends on the choice of c.

Hence it is enough to bound the contribution of small frequencies, $0 < |S| < c\log n$, for some constant c that will be chosen later. As previously we have for any $\rho \in (0,1)$ and using hypercontractivity,

$$\sum_k \sum_{0<|S|<c\log n} \widehat{\nabla_k f}(S)^2 \leq \rho^{-2c\log n} \sum_k \|T_\rho \nabla_k f\|_2^2$$

$$\leq \rho^{-2c\log n} \sum_k \|\nabla_k f\|_{1+\rho^2}^2$$

$$\leq \rho^{-2c\log n} \sum_k I_k(f)^{2/(1+\rho^2)}$$

$$\leq \rho^{-2c\log n}(\sup_k I_k(f))^{\frac{1-\rho^2}{1+\rho^2}} \sum_k I_k(f)$$

$$\leq \rho^{-2c\log n} n^{-\alpha\frac{1-\rho^2}{1+\rho^2}} \sum_k I_k(f) \text{ by our assumption}.$$

$$(7.1)$$

Now fixing any $\rho \in (0,1)$, and then choosing the constant c depending on ρ and α, the lemma follows. By optimizing on the choice of ρ, one could get better constants if one wants.   □

7.5 Further discussion

Some words on the proof of Theorem 7.2

The main difficulty here is that the quantity of interest, $f(\omega) := \text{dist}_\omega(0,v)$, is no longer invariant under a large class of graph automorphisms. This lack of symmetry makes the study of influences more difficult. For example, edges near the endpoints 0 or v have very high influence (of order 1). To gain some more symmetry, the authors in (BKS03) rely on a very nice "averaging" procedure. We refer to this paper for more details.

Known lower bounds on the fluctuations

We discussed mainly here ways to obtain upper bounds on the fluctuations of the shapes in first-passage percolation. It is worth pointing out that some nontrivial *lower* bounds on the fluctuations are known for $\mathbb{Z}^2$. See (PP94; NP95).

Remark We end by mentioning that the proof given in (BKS03) was based on an inequality by Talagrand. The proof given here avoids this inequality.

7.6 Exercises

7.1 Let $n \geq 1$ and $d \geq 2$. Consider the random metric on the torus $\mathbb{Z}^d / n\mathbb{Z}^d$ as described in this chapter. For any $k \geq 1$, let $\mathcal{A}_n^k$ be the event that the shortest "horizontal" circuit is $\leq k$. If $d \geq 3$, show that for any choice of $k_n = k(n)$, the family of events $\mathcal{A}_n^{k_n}$ is noise sensitive. (Note that the situation here is similar to the Problem 1.9 in Chapter 1.) Finally, discuss the two-dimensional case, $d = 2$ (nonrigorously).

7.2 Show that Lemma 7.4 is false if $\mathbf{I}_k(f)$ is taken to be the square of the L^2 norm of $\nabla_k f$ rather than the probability of its support (i.e., find a counterexample).

8

Randomized algorithms and noise sensitivity

In this chapter, we explain how the notion of **revealment** for so-called randomized algorithms can in some cases yield direct information concerning the energy spectrum that may allow not only noise sensitivity results but even quantitative noise sensitivity results.

8.1 BKS and randomized algorithms

In the previous chapter, we explained how Theorem 1.21 together with bounds on the pivotal exponent for percolation yields noise sensitivity for percolation crossings. However, in (BKS99), a different approach was in fact used for showing noise sensitivity which, while still using Theorem 1.21, did not use these bounds on the critical exponent. In that approach, one sees the first appearance of randomized algorithms. In a nutshell, the authors showed that (1) if a monotone function is very uncorrelated with all majority functions, then it is noise sensitive (in a precise quantitative sense) and (2) percolation crossings are very uncorrelated with all majority functions. The latter is shown by constructing a certain algorithm which, due to the RSW Theorem 2.1, looks at very few bits but still looks at enough bits to be able to determine the output of the function. This approach is detailed in Section 12.13 in Chapter 12.

8.2 The revealment theorem

An **algorithm** for a Boolean function f is an algorithm A that queries (asks the values of) the bits one by one, where the decision of which bit to ask can be based on the values of the bits previously queried, and stops once f is determined (being determined means that f takes the same value no matter how the remaining bits are set).

A **randomized algorithm** for a Boolean function f is the same as above but auxiliary randomness may also be used to decide the next value queried (including for the first bit). [In computer science, the term randomized decision tree would be used for our notion of randomized algorithm, but we will not use this terminology.]

The following definition of *revealment* will be crucial. Given a randomized algorithm A for a Boolean function f, we let J_A denote the random set of bits queried by A. (Note that this set depends both on the randomness corresponding to the choice of ω and the randomness inherent in running the algorithm, which are of course taken to be independent.)

Definition 8.1 The **revealment of a randomized algorithm** A for a Boolean function f, denoted by δ_A, is defined by

$$\delta_A := \max_{i \in [n]} \mathbb{P}(i \in J_A).$$

The **revealment of a Boolean function** f, denoted by δ_f, is defined by

$$\delta_f := \inf_A \delta_A$$

where the infimum is taken over all randomized algorithms A for f.

This section presents a connection between noise sensitivity and randomized algorithms. It will be used later to yield an alternative proof of noise sensitivity for percolation crossings that is not based on Theorem 1.21 (or Proposition 5.6). Two other advantages of the algorithmic approach of the present section over that mentioned in the previous section (besides the fact that it does not rest on Theorem 1.21) is that it applies to nonmonotone functions and yields a more "quantitative" version of noise sensitivity.

We have defined algorithms, randomized algorithms, and revealment only for Boolean functions but the definitions immediately extend to functions $f : \Omega_n \to \mathbb{R}$.

The main theorem of this section is the following.

Theorem 8.2 (SS10) *For any function* $f : \Omega_n \to \mathbb{R}$ *and for each* $k = 1, 2, \ldots$, *we have that*

$$E_f(k) = \sum_{S \subseteq [n], |S| = k} \hat{f}(S)^2 \leq \delta_f k \|f\|^2, \tag{8.1}$$

where $\|f\|$ *denotes the* L^2 *norm of* f *with respect to the uniform probability measure on* Ω *and* δ_f *is the revealment of* f.

Before giving the proof, we make some comments to help readers see what is happening and suggest why a result like this might be true. Our original function is a sum of monomials with coefficients given by the Fourier coefficients. Each time a bit is revealed by the algorithm, we obtain a new Boolean function obtained by just substituting in the value of the bit we obtained into the corresponding variable. On the algebraic side, those monomials that contain this bit go down by 1 in degree while the other monomials are unchanged. There might, however, be cancellation in the process, which is what we hope for because when the algorithm stops, all the monomials (except the constant) must have been killed. The way cancelation occurs is illustrated as follows. The Boolean function at some stage might contain $(1/3)x_2x_4x_5 + (1/3)x_2x_4$ and then the bit x_5 might be revealed and take the value -1. When we substitute this value into the variable, the two terms cancel and disappear, thereby bringing us one step closer to a constant (and hence determined) function.

As far as why the result might be true, the intuition, very roughly speaking, is as follows. The theorem says that for a Boolean function we cannot, for example, have $\delta = 1/1000$ and $\sum_i \hat{f}(\{i\})^2 = 1/2$. If the level 1 monomials of the function were

$$a_1\omega_1 + a_2\omega_2 + \cdots + a_n\omega_n,$$

then it is clear that after the algorithm is over, then with high probability, the sum of the squares of the coefficients of the terms that have not been reduced to a constant is still reasonably large. Therefore, because the function at the end of the algorithm is constant, these remaining terms must necessarily have been canceled by higher degree monomials which, after running the algorithm, have been "reduced to" degree 1 monomials. If, for the sake of this heuristic argument, we assume that each bit is revealed independently, then the probability that a degree $k \geq 2$ monomial is brought down to a degree 1 monomial (which is necessary for it to help to cancel the degree 1 terms described above) is at most δ^{k-1} and hence the expected sum of the squares of the coefficients from the degree $k \geq 2$ monomials that are brought down to degree 1 is at most δ^{k-1}. The total such sum for levels 2 to n is then at most

$$\sum_{k=2}^{n} \delta^{k-1} \leq 2\delta$$

which won't be enough to cancel the (originally) degree 1 monomials that remained degree 1 after running the algorithm if δ is much less than $\sum_i \hat{f}(\{i\})^2$. A similar heuristic works for the other levels.

Proof In the following, we let $\tilde{\Omega}$ denote the probability space that includes the randomness in the input bits of f and the randomness used to run the algorithm (which we assume to be independent) and we let $\mathbb{E}$ denote the corresponding expectation. Without loss of generality, elements of $\tilde{\Omega}$ can be represented as $\tilde{\omega} = (\omega, \tau)$ where ω are the random bits and τ represents the randomness necessary to run the algorithm.

Now, fix $k \geq 1$. Let

$$g(\omega) := \sum_{|S|=k} \hat{f}(S) \chi_S(\omega), \qquad \omega \in \Omega.$$

The left-hand side of (8.1) is equal to $\|g\|^2$.

Let $J \subseteq [n]$ be the random set of all bits examined by the algorithm. Let $\mathcal{A}$ denote the minimal σ-field for which J is measurable and every ω_i, $i \in J$, is measurable; this can be viewed as the relevant information gathered by the algorithm. For any function $h \colon \Omega \to \mathbb{R}$, let $h_J \colon \Omega \to \mathbb{R}$ denote the random function obtained by substituting the values of the bits in J. More precisely, if $\tilde{\omega} = (\omega, \tau)$ and $\omega' \in \Omega$, then $h_J(\tilde{\omega})(\omega')$ is $h(\omega'')$ where ω'' is ω on $J(\tilde{\omega})$ and is ω' on $[n] \setminus J(\tilde{\omega})$. In this way, h_J is a random variable on $\tilde{\Omega}$ taking values in the set of mappings from Ω to $\mathbb{R}$ and it is immediate that this random variable is $\mathcal{A}$-measurable. When the algorithm terminates, the unexamined bits in Ω are unbiased and hence $\mathbb{E}[h|\mathcal{A}] = \int h_J (= \hat{h}_J(\emptyset))$ where $\int$ is defined, as usual, to be integration with respect to uniform measure on Ω. It follows that $\mathbb{E}[h] = \mathbb{E}[\int h_J]$.

Similarly, for all h,

$$\|h\|^2 = \mathbb{E}[h^2] = \mathbb{E}\left[\int h_J^2\right] = \mathbb{E}[\|h_J\|^2]. \tag{8.2}$$

Because the algorithm determines f, it is $\mathcal{A}$ measurable, and we have

$$\|g\|^2 = \mathbb{E}[g\,f] = \mathbb{E}\big[\mathbb{E}[g\,f\,|\mathcal{A}]\big] = \mathbb{E}\big[f\,\mathbb{E}[g|\mathcal{A}]\big].$$

Since $\mathbb{E}[g|\mathcal{A}] = \hat{g}_J(\emptyset)$, Cauchy–Schwarz therefore gives

$$\|g\|^2 \leq \sqrt{\mathbb{E}[\hat{g}_J(\emptyset)^2]} \|f\|. \tag{8.3}$$

We now apply Parseval's formula to the (random) function g_J: this gives (for any $\tilde{\omega} = (\omega, \tau) \in \tilde{\Omega}$),

$$\hat{g}_J(\emptyset)^2 = \|g_J\|_2^2 - \sum_{|S|>0} \hat{g}_J(S)^2.$$

Taking the expectation over $\tilde{\omega} \in \tilde{\Omega}$, this leads to

$$\mathbb{E}[\hat{g}_J(\emptyset)^2] = \mathbb{E}[\|g_J\|_2^2] - \sum_{|S|>0} \mathbb{E}[\hat{g}_J(S)^2]$$

$$= \|g\|_2^2 - \sum_{|S|>0} \mathbb{E}[\hat{g}_J(S)^2] \quad \text{by (8.2)}$$

$$= \sum_{|S|=k} \hat{g}(S)^2 - \sum_{|S|>0} \mathbb{E}[\hat{g}_J(S)^2] \left\{ \begin{array}{l} \text{since } g \text{ is supported} \\ \text{on level-}k \text{ coefficients} \end{array} \right.$$

$$\leq \sum_{|S|=k} \mathbb{E}[\hat{g}(S)^2 - \hat{g}_J(S)^2] \left\{ \begin{array}{l} \text{by restricting to} \\ \text{level-}k \text{ coefficients} \end{array} \right.$$

Now, since g_J is built randomly from g by fixing the variables in $J = J(\tilde{\omega})$, and since g by definition does not have frequencies larger than k, it is clear that for any S with $|S| = k$ we have

$$\hat{g}_J(S) = \left\{ \begin{array}{ll} \hat{g}(S) = \hat{f}(S), & \text{if } S \cap J(\tilde{\omega}) = \emptyset \\ 0, & \text{otherwise.} \end{array} \right.$$

Therefore, we obtain

$$\|\mathbb{E}[g \mid J]\|_2^2 = \mathbb{E}[\hat{g}_J(\emptyset)^2] \leq \sum_{|S|=k} \hat{g}(S)^2 \, \mathbb{P}[S \cap J \neq \emptyset] \leq \|g\|_2^2 \, k\delta.$$

Combining with (8.3) completes the proof. $\qquad\qquad\square$

Proposition 4.2 and Theorem 8.2 immediately imply the following corollary.

Corollary 8.3 *If the revealments satisfy*

$$\lim_{n \to \infty} \delta_{f_n} = 0,$$

then $\{f_n\}$ is noise sensitive.

In the exercises, one is asked to show that certain sequences of Boolean functions are noise sensitive by applying the preceding corollary.

8.3 An application to noise sensitivity of percolation

In this section, we apply Corollary 8.3 to prove noise sensitivity of percolation crossings. The following result gives the necessary assumption that the revealments approach 0.

Theorem 8.4 ((SS10)) *Let $f = f_n$ be the indicator function for the event that critical site percolation on the triangular grid contains a left to right crossing of our $n \times n$ box. Then $\delta_{f_n} \leq n^{-1/4+o(1)}$ as $n \to \infty$.*

For critical bond percolation on the square grid, this holds with $1/4$ replaced by some positive constant $a > 0$.

Outline of Proof. We outline the argument only for the triangular lattice; the argument for the square lattice is similar. We begin by giving a first attempt at a good algorithm. We consider from Chapter 2 the exploration path or interface from the bottom right of the square to the top left used to detect a left–right crossing. This (deterministic) algorithm simply asks the bits that it needs to know in order to continue the interface. Observe that if a bit is queried, it is necessarily the case that there is both a black and white path from next to the hexagon to the boundary. It follows, from the exponent of $1/4$ for the 2-arm event in Chapter 2, that, for hexagons far from the boundary, the probability that they are revealed is at most $R^{-1/4+o(1)}$, as desired. However, one cannot conclude that points near the boundary have small revealment and of course the right bottom point is always revealed.

The way that we modify the above algorithm so that all points have small revealment is as follows. We first choose a point x at random from the middle third of the right side. We then run two algorithms, the first one which checks whether there is a left–right path from the right side *above* x to the left side and the second one which checks whether there is a left–right path from the right side *below* x to the left side. The first part is done by looking at an interface from x to the top left corner as above. The second part is done by looking at an interface from x to the bottom left corner as above (but where the colors on the two sides of the interface need to be swapped). See Figure 8.1.

It can then be shown with a little work (but no new conceptual ideas) that this modified algorithm has the desired revealment of at most $R^{-1/4+o(1)}$ as desired. One of the things that one needs to use in this analysis is the so-called one-arm half-plane exponent, which has a known value of $1/3$. See (SS10) for details. □

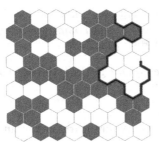 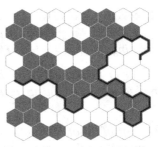

Figure 8.1 Illustration of the algorithm to determine if there is a left-right crossing

8.3.1 First quantitative noise sensitivity result

In this subsection, we give our first "polynomial bound" on the noise sensitivity of percolation. This is an important step in our understanding of quantitative noise sensitivity of percolation initiated in Chapter 6.

Recall that in the definition of noise sensitivity, ϵ is held fixed. However, as we have seen in Chapter 6, it is of interest to ask if the correlations can still go to 0 when $\epsilon = \epsilon_n$ goes to 0 with n but not so fast. The techniques of the present chapter imply the following result.

Theorem 8.5 (SS10) *Let $\{f_n\}$ be as in Theorem 8.4. Then, for the triangular lattice, for all $\gamma < 1/8$,*

$$\lim_{n \to \infty} \mathbb{E}[f_n(\omega)f_n(\omega_{1/n^\gamma})] - \mathbb{E}[f_n(\omega)]^2 = 0. \tag{8.4}$$

On the square lattice, there exists some $\gamma > 0$ with the above property.

Proof   We prove only the first statement; the square lattice case is handled similarly. First, (4.3) gives us that every n and γ,

$$\mathbb{E}[f_n(\omega)f_n(\omega_{1/n^\gamma})] - \mathbb{E}[f_n(\omega)]^2 = \sum_{k=1} E_{f_n}(k)(1 - 1/n^\gamma)^k. \tag{8.5}$$

Note that there are order n^2 terms in the sum. Fix $\gamma < 1/8$. Choose $\epsilon > 0$ so that $\gamma + \epsilon < 1/8$. For large n, we have that $\delta_{f_n} \leq 1/n^{1/4-\epsilon}$. The right-hand side of (8.5) is at most

$$\sum_{k=1}^{n^{\gamma+\epsilon/2}} k/n^{1/4-\epsilon} + (1 - 1/n^\gamma)^{n^{\gamma+\epsilon/2}}$$

by breaking up the sum at $n^{\gamma+\epsilon/2}$ and applying Theorems 8.2 and 8.4 to bound the $E_{f_n}(k)$ terms in the first part. The second term clearly goes to 0 while the first part also goes to 0 by the way ϵ was chosen. □

Observe that the *proof* of Theorem 8.5 immediately yields the following general result.

Corollary 8.6 *Let* $\{f_n\}$ *be a sequence of Boolean functions on* m_n *bits with* $\delta(f_n) \leq O(1)/n^\beta$ *for all* n. *Then for all* $\gamma < \beta/2$, *we have that*

$$\lim_{n\to\infty} \mathbb{E}[f_n(\omega)f_n(\omega_{1/n^\gamma})] - \mathbb{E}[f_n(\omega)]^2 = 0. \tag{8.6}$$

8.4 Lower bounds on revealments

One of the goals of the present section is to show that one cannot hope to reach the conjectured 3/4-sensitivity exponent with Theorem 8.2. Theorem 8.5 told us that we obtain asymptotic decorrelation if the noise is $1/n^\gamma$ for $\gamma < 1/8$. Note that this differs from the conjectured "critical exponent" of 3/4 by a factor of 6. In this section, we investigate the degree to which the 1/8 could potentially be improved and in the discussion, we bring up an interesting open problem. We also derive an interesting general theorem giving a nontrivial lower bound on the revealment for monotone functions. We start with the following definition.

Definition 8.7 Given a randomized algorithm A for a Boolean function f, let $C(A)$ (the cost of A) be the expected number of queries that the algorithm A makes. Let $C(f)$ (the cost of f) be the infimum of $C(A)$ over all randomized algorithms A for f.

Remarks (1). It is easy to see that $C(f)$ is unchanged if we take the infimum over deterministic algorithms.
(2). Clearly $n\delta_A \geq C(A)$ and hence $n\delta_f \geq C(f)$.
(3). $C(f)$ is at least the total influence $\mathbf{I}(f)$ since for any algorithm A and any i, the event that i is pivotal necessarily implies that the bit i is queried by A.

The following result due to O'Donnell and Servedio (OS07) is an essential improvement on the third part of the last remark.

Theorem 8.8 *Let* f *be a monotone Boolean function mapping* Ω_n *into* $\{-1, 1\}$. *Then* $C(f) \geq \mathbf{I}(f)^2$ *and hence* $\delta_f \geq \mathbf{I}(f)^2/n$.

Proof Fix any randomized algorithm A for f. Let $J = J_A$ be the random set of bits queried by A. We then have

$$\mathbf{I}(f) = \mathbb{E}[\sum_i f(\omega)\omega_i] = \mathbb{E}[f(\omega)\sum_i \omega_i I_{\{i\in J\}}] \le \sqrt{\mathbb{E}[f(\omega)^2]}\sqrt{\mathbb{E}[(\sum_i \omega_i I_{\{i\in J\}})^2]}$$

where the first equality uses monotonicity (recall Proposition 4.5) and then the Cauchy–Schwarz inequality is used. We now bound the first term by 1. For the second moment inside the second square root, the sum of the diagonal terms yields $\mathbb{E}[|J|]$ while the cross terms are all 0 since for $i \ne j$, $\mathbb{E}[\omega_i I_{\{i\in J\}}\omega_j I_{\{j\in J\}}] = 0$ as can be seen by breaking up the sum depending on whether i or j is queried first. This yields the result. □

Returning to our event f_n of percolation crossings, because the sum of the influences is $n^{3/4+o(1)}$, Theorem 8.8 tells us that $\delta_{f_n} \ge n^{-1/2+o(1)}$. It follows from the method of proof in Theorem 8.5 that Theorem 8.2 cannot improve the result of Theorem 8.5 past $\gamma = 1/4$ which is still a factor of 3 from the critical value $3/4$. Of course, one could investigate the degree to which Theorem 8.2 itself could be improved.

Theorem 8.4 tells us that there are algorithms A_n for f_n such that $C(A_n) \le n^{7/4+o(1)}$. On the other hand, Theorem 8.8 tell us that it is necessarily the case that $C(A) \ge n^{6/4+o(1)}$.

Open Question: Find the smallest σ such that there are algorithms A_n for f_n with $C(A_n) \le n^\sigma$. (We know $\sigma \in [6/4, 7/4]$.)

We mention another inequality relating revealment with influences that is a consequence of the results in (OSSS05). One of the main results from this paper is proved in and is the highlight of Section 12.10 in Chapter 12, which also describes the concepts of deterministic and randomized complexity.

Theorem 8.9 *Let f be a Boolean function mapping Ω_n into $\{-1, 1\}$. Then*
$$\delta_f \ge \mathrm{Var}(f)/(n \max_i \mathbf{I}_i(f))$$

It is interesting to compare Theorems 8.8 and 8.9. Assuming $\mathrm{Var}(f)$ is of order 1, and all the influences are of order $1/n^\alpha$, then it is easy to check that Theorem 8.8 gives a better bound when $\alpha < 2/3$ and Theorem 8.9 gives a better bound when $\alpha > 2/3$. For crossings of percolation, where α should be $5/8$, it is better to use Theorem 8.8 rather than 8.9.

Finally, there are a number of interesting results concerning revealment obtained in the paper (BSW05). Four results are as follows.
(i) If f is reasonably balanced on n bits, then the revealment is at least of order $1/n^{1/2}$.

(ii) There is a reasonably balanced function on n bits whose revealment is at most $O(1)(\log n)/n^{1/2}$.

(iii) If f is reasonably balanced on n bits and is monotone, then the revealment is at least of order $1/n^{1/3}$.

(iv) There is a reasonably balanced monotone function on n bits whose revealment is at most $O(1)(\log n)/n^{1/3}$.

We finally end this section by giving one more reference that gives an interesting connection between percolation, algorithms, and game theory; see (PSSW07).

8.5 An application to a critical exponent

In this section, we show how Theorem 8.2 or in fact Theorem 8.8 can be used to prove Proposition 6.6, which says that the four-arm exponent is strictly larger than 1. As we mentioned earlier, this can be shown for the triangular lattice using SLE technology. The following proof of Proposition 6.6 works for both $\mathbb{T}$ and on $\mathbb{Z}^2$.

We will assume the separation of arms result mentioned earlier in Chapter 6 which says that for the event f_R, the influence of any variable further than distance $R/10$ from the boundary is $\asymp \alpha_4(R)$.

Proof of Proposition 6.6. Theorems 8.2 and 8.4 imply that for some $a > 0$,

$$\sum_i \hat{f}_R(\{i\})^2 \leq 1/R^a.$$

Next, denote the set of variables further than distance $R/10$ from the boundary by B (for bulk). Using the separation of arms as explained previously, we have

$$R^2 \alpha_4^2(R) \leq O(1) \sum_{i \in B} \mathbf{I}_i^2. \tag{8.7}$$

Proposition 4.5 then yields

$$R^2 \alpha_4^2(R) \leq O(1/R^a)$$

and the result follows. □

Observe that Theorem 8.8 could also be used as follows. Theorem 8.4 implies that $C(f_R) \leq R^{2-a}$ for some $a > 0$ and then Theorem 8.8 yields $\mathbf{I}(f_R)^2 \leq R^{2-a}$.

Exactly as in (8.7), one has, again using separation of arms, that

$$R^2 \alpha_4(R) \leq O(1) \sum_{i \in B} \mathbf{I}_i \leq O(1)\mathbf{I}(f_R). \tag{8.8}$$

Altogether this gives us

$$R^4 \alpha_4^2(R) \le O(1)R^{2-a},$$

again yielding the result.

We finally mention that it is not so strange that either of Theorems 8.2 or 8.8 can be used here because, as readers can easily verify, for the case of monotone functions all of whose variables have the same influence, the case $k = 1$ in Theorem 8.2 is equivalent to Theorem 8.8.

Remark (1) We mention here that the proof for the multiscale version of Proposition 6.6 in the appendix of (SS11) is an extension of the approach of O'Donnell and Servedio above.

(2) See Theorem 12.40 at the end of Section 12.10 in the miscellaneous Chapter 12 for another example of a nontrivial inequality about arms-events obtained using randomized algorithms ideas.

8.6 Does noise sensitivity imply low revealment?

As far as this book is concerned, this subsection does not connect to anything that follows and hence can be viewed as tangential.

It is natural to ask if the converse of Corollary 8.3 might be true. A moment's thought reveals that Example 1.3, Parity, provides a counterexample. However, it is more interesting perhaps that there is a monotone counterexample to the converse which is provided by Example 1.6, Clique containment.

Proposition 8.10 *Clique containment provides an example showing that the converse of Corollary 8.3 is false for monotone functions.*

Outline of Proof. We first explain more precisely the size of the clique that we are looking for. Given n and k, let $f(n,k) := \binom{n}{k} 2^{-\binom{k}{2}}$, which is just the expected number of cliques of size k in a random graph. When k is around $2\log_2(n)$, it is easy to check that $f(n, k+1)/f(n, k)$ is $o(1)$ as $n \to \infty$. For such k, clearly if $f(n,k)$ is small, then with high probability there is no k-clique while it can be shown, via a second moment type argument, that if $f(n,k)$ is large, then with high probability there is a k-clique. One now takes k_n to be around $2\log_2(n)$ such that $f(n, k_n) \ge 1$ and $f(n, k_n + 1) < 1$. Since $f(n, k+1)/f(n, k)$ is $o(1)$, it follows with some thought from the above that the clique number is concentrated on at most two points. Furthermore, if $f(n, k_n)$ is very large and $f(n, k_n + 1)$ very small, then it is concentrated on one point. Again, see (AS00) for details.

Finally, we denote the event that the random graph on n vertices contains a clique of size k_n by A_n. We have already seen in one of the exercises that this example is noise sensitive. We consider only a sequence of n's so that A_n is nondegenerate in the sense that the probabilities of this sequence stay bounded away from 0 and 1. An interesting point is that there is such a sequence. Again, see (AS00) for this. To show that the revealments do not go to 0, it suffices to show that the sequence of costs (see Definition 8.7 and the remarks afterwards) is $\Omega(n^2)$. We prove something stronger but, to do this, we must first give a few more definitions.

Definition 8.11 For a given Boolean function f, a **witness** for ω is any subset W of the variables such that the elements of ω in W determine f in the sense that for every ω' which agrees with ω on W, we have that $f(\omega) = f(\omega')$. The **witness size** of ω, denoted $w(\omega)$, is the size of the smallest witness for ω. The **expected witness size**, denoted by $w(f)$, is $\mathbb{E}(w(\omega))$.

Observe that, for any Boolean function f, the bits revealed by any algorithm A for f and for any ω is always a witness for ω. It easily follows that the cost $C(f)$ satisfies $C(f) \geq w(f)$. Therefore, to prove the proposition, it suffices to show that

$$w(f_n) = \Omega(n^2). \tag{8.9}$$

Remark
(1) The above also implies that with a fixed uniform probability, $w(\omega)$ is $\Omega(n^2)$.
(2) Of course when f_n is 1, there is always a (small) witness of size $\binom{k_n}{2} \ll n$ and so the large average witness size comes from when f_n is -1.
(3) However, it is not deterministically true that when f_n is -1, $w(\omega)$ is necessarily of size $\Omega(n^2)$. For example, for $\omega \equiv -1$ (corresponding to the empty graph), the witness size is $o(n^2)$ as is easily checked. Clearly the empty graph has the smallest witness size among ω with $f_n = -1$.

Lemma 8.12 *Let E_n be the event that all sets of vertices of size at least $.97n$ contains C_{k_n-3}. Then $\lim_{n \to \infty} \mathbb{P}(E_n) = 1$.*

Proof This follows, after some work, from the Janson inequalities. See (AS00) for details concerning these inequalities. □

Lemma 8.13 *Let U be any collection of at most $n^2/1000$ edges in C_n. Then there exist distinct v_1, v_2, v_3 such that no edge in U goes between any v_i and v_j and*

$$|\{e \in U : e \text{ is an edge between } \{v_1, v_2, v_3\} \text{ and } \{v_1, v_2, v_3\}^c\}| \leq n/50. \tag{8.10}$$

Proof   We use the probabilistic method where we choose $\{v_1, v_2, v_3\}$ to be a uniformly chosen 3-set. It is immediate that the probability that the first condition fails is at most $3|U|/\binom{n}{2} \le 1/100$. Letting Y be the number of edges in the set appearing in (8.10) and Y' be the number of U edges touching v_1, it is easy to see that

$$\mathbb{E}(Y) \le 3\mathbb{E}(Y') = 6|U|/n \le n/100$$

where the equality follows from the fact that, for any graph, the number of edges is half the total degree. By Markov's inequality, the probability of the event in (8.10) holds with probably at least $1/2$. This shows that the random 3-set $\{v_1, v_2, v_3\}$ satisfies the two stated conditions with positive probability and hence such a 3-set exists.   □

By Lemma 8.12, we have $\mathbb{P}(A_n^c \cap E_n) \ge c > 0$ for all large n. To prove the theorem, it therefore suffices to show that if $A_n^c \cap E_n$ occurs, there is no witness of size smaller than $n^2/1000$. Assume U to be any set of edges of size smaller than $n^2/1000$. Choose $\{v_1, v_2, v_3\}$ from Lemma 8.13. By the second condition in this lemma, there exists a set S of size at least $0.97n$ that is disjoint from $\{v_1, v_2, v_3\}$ that has no U-edge to $\{v_1, v_2, v_3\}$. Since E_n occurs, S contains a C_{k_n-3}, whose vertices we denote by T. Because there are no U-edges between T and $\{v_1, v_2, v_3\}$ or within $\{v_1, v_2, v_3\}$ (by the first condition in Lemma 8.13) and T is the complete graph, U cannot be a witness since A_n^c occured.   □

The key step in the proof of Proposition 8.10 is (8.9). This is stated without proof in (FKW02); however, E. Friedgut provided us with the above proof.

8.7 Exercises

8.1   Compute the revealment for Majority function on 3 bits.

8.2   Use Corollary 8.3 to show that Examples 1.5 and 1.15, Iterated 3-Majority function and Tribes, are noise sensitive.

8.3   For transitive monotone functions, is there a relationship between revealment and the minimal cost over all algorithms?

8.4   Show that for transitive monotone functions, Theorem 8.8 yields the same result as Theorem 8.2 does for the case $k = 1$.

8.5   What can you say about the sequence of revealments for the Iterated

3-Majority function? [It can be shown that the sequence of revealments decays like $1/n^\sigma$ for some σ but it is an open question what σ is.]

8.6 You are given a sequence of Boolean functions and told that it is not noise sensitive using noise $\epsilon_n = 1/n^{1/5}$. What, if anything, can you conclude about the sequence of revealments δ_n?

8.7 Note that a consequence of Corollary 8.3 and the converse to Theorem 1.21 for monotone functions is that if $\{f_n\}$ is a sequence of monotone functions, then, if the revealments of $\{f_n\}$ go to 0, then the sums of the squared influences approach 0. Show that this implication is false without the monotonicity assumption.

9

The spectral sample

It turns out that it is very useful to view the Fourier coefficients of a Boolean function as a random subset of the input bits where the "weight" or "probability" of a subset is its squared Fourier coefficient. It is our understanding that it was Gil Kalai who suggested that thinking of the spectrum as a random set could shed some light on the types of questions we are looking at here. The following is the crucial definition in this chapter.

9.1 Definition of the spectral sample

Definition 9.1 Given a Boolean function $f: \Omega_n \to \{\pm 1\}$ or $\{0, 1\}$, we let the **spectral measure** $\hat{\mathbb{Q}} = \hat{\mathbb{Q}}_f$ of f be the measure on subsets $\{1, \ldots, n\}$ given by

$$\hat{\mathbb{Q}}_f(S) := \hat{f}(S)^2, \ S \subset \{1, \ldots, n\}.$$

We let $\mathscr{S}_f = \mathscr{S}$ denote a subset of $\{1, \ldots, n\}$ chosen according to this measure and call this the **spectral sample**. We let $\hat{\mathbb{Q}}$ also denote the corresponding expectation (even when $\hat{\mathbb{Q}}$ is not a probability measure).

By Parseval's formula, the total mass of the so-defined spectral measure is

$$\sum_{S \subset \{1, \ldots, n\}} \hat{f}(S)^2 = \mathbb{E}[f^2].$$

This makes the following definition natural.

Definition 9.2 Given a Boolean function $f: \Omega_n \to \{\pm 1\}$ or $\{0, 1\}$, we let the **spectral probability measure** $\hat{\mathbb{P}} = \hat{\mathbb{P}}_f$ of f be the probability measure on subsets of $\{1, \ldots, n\}$ given by

$$\hat{\mathbb{P}}_f(S) := \frac{\hat{f}(S)^2}{\mathbb{E}[f^2]}, \ S \subset \{1, \ldots, n\}.$$

Since $\hat{\mathbb{P}}_f$ is just $\hat{\mathbb{Q}}_f$ up to a renormalization factor, the **spectral sample**

$\mathscr{S}_f = \mathscr{S}$ will denote as well a random subset of $[n]$ sampled according to $\hat{\mathbb{P}}_f$. We let $\hat{\mathbb{E}}_f = \hat{\mathbb{E}}$ denote its corresponding expectation.

Remark (1) Note that if f maps into $\{\pm 1\}$, then, by Parseval's formula, $\hat{\mathbb{Q}}_f = \hat{\mathbb{P}}_f$ while if it maps into $\{0, 1\}$, $\hat{\mathbb{Q}}_f$ will be a subprobability measure.
(2) Observe that if $(f_n)_n$ is a sequence of nondegenerate Boolean functions into $\{0, 1\}$, then $\hat{\mathbb{P}}_{f_n} \asymp \hat{\mathbb{Q}}_{f_n}$.
(3) There is no statistical relationship between ω and $\mathscr{S}_f$ as they are defined on different probability spaces. The spectral sample will just be a convenient point of view in order to understand the questions we are studying.

Some of the formulas and results we have previously derived in earlier chapters have very simple formulations in terms of the spectral sample. For example, it is immediate to check that (4.2) simply becomes

$$\mathbb{E}[f(\omega)f(\omega_\epsilon)] = \hat{\mathbb{Q}}_f[(1 - \epsilon)^{|\mathscr{S}|}] \tag{9.1}$$

or

$$\mathbb{E}[f(\omega)f(\omega_\epsilon)] - \mathbb{E}[f(\omega)]^2 = \hat{\mathbb{Q}}_f[(1 - \epsilon)^{|\mathscr{S}|} I_{\mathscr{S} \neq \emptyset}]. \tag{9.2}$$

Next, in terms of the spectral sample, Propositions 4.2 and 4.3 simply become the following proposition.

Proposition 9.3 *If $\{f_n\}$ is a sequence of Boolean functions mapping into $\{\pm 1\}$, then we have the following.*
(i) $\{f_n\}$ is noise sensitive if and only if $|\mathscr{S}_{f_n}| \to \infty$ in probability on the set $\{|\mathscr{S}_{f_n}| \neq 0\}$.
(ii) $\{f_n\}$ is noise stable if and only if the random variables $\{|\mathscr{S}_{f_n}|\}$ are tight.

There is also a nice relationship between the pivotal set $\mathcal{P}$ and the spectral sample. The following result, which is simply Proposition 4.4 (see also the remark after this proposition), tells us that the two random sets $\mathcal{P}$ and $\mathscr{S}$ have the same one-dimensional marginals.

Proposition 9.4 *If f is a Boolean function mapping into $\{\pm 1\}$, then for all $i \in [n]$ we have that*

$$\mathbb{P}(i \in \mathcal{P}) = \hat{\mathbb{Q}}(i \in \mathscr{S})$$

and hence $\mathbb{E}(|\mathcal{P}|) = \hat{\mathbb{Q}}(|\mathscr{S}|)$.

(This proposition is stated with $\hat{\mathbb{Q}}$ instead of $\hat{\mathbb{P}}$ since if f maps into $\{0, 1\}$ instead, then readers can check that the above holds with an extra factor of 4 on the right-hand side while if $\hat{\mathbb{P}}$ were used instead, then this would not be true for any constant.) Even though $\mathscr{S}$ and $\mathcal{P}$ have the same "one-dimensional" marginals, it is not, however, true that these two random sets have the same distribution. For example, it is easily checked that for $\mathbf{MAJ_3}$, these two distributions are different. Interestingly, as we will see in the next section, $\mathscr{S}$ and $\mathcal{P}$ also always have the same "two-dimensional" marginals. This will prove useful when applying second moment method arguments.

Before ending this section, let us give an alternative proof of Proposition 6.12 using this point of view of thinking of $\mathscr{S}$ as a random set.

Alternative proof of Proposition 6.12

The statement of the proposition when converted to the spectral sample states (see the exercises in this chapter if this is not clear) that for any $a_n \to \infty$,

$$\lim_{n \to \infty} \hat{\mathbb{P}}(|\mathscr{S}_n| \geq a_n n^2 \alpha_4(n)) = 0.$$

However this immediately follows from Markov's inequality using Propositions 6.8 and 9.4.   □

9.2 A way to sample the spectral sample in a sub-domain

In this section, we describe a method of "sampling" the spectral measure restricted to a subset of the bits. As an application of this, we show that $\mathscr{S}$ and $\mathcal{P}$ in fact have the same two-dimensional marginals, namely that for all i and j, $\mathbb{P}(i, j \in \mathcal{P}) = \hat{\mathbb{Q}}(i, j \in \mathscr{S})$.

To first get a little intuition about the spectral measure, we start with an easy proposition.

Proposition 9.5 (GPS10)   *For a Boolean function f and $A \subseteq \{1, 2, \ldots, n\}$, we have*

$$\hat{\mathbb{Q}}(\mathscr{S}_f \subseteq A) = \mathbb{E}[|\mathbb{E}(f|A)|^2]$$

where conditioning on A means conditioning on the bits in A.

Proof   Noting that $\mathbb{E}(\chi_S|A)$ is χ_S if $S \subseteq A$ and 0 otherwise, we obtain by expanding that

$$\mathbb{E}(f|A) = \sum_{S \subseteq A} \hat{f}(S) \chi_S.$$

Now apply Parseval's formula.   □

If we have a subset $A \subseteq \{1, 2, \ldots, n\}$, how do we "sample" from $A \cap \mathscr{S}$? A nice way to proceed is as follows: choose a random configuration outside of A, then look at the induced function on A and sample from the induced function's spectral measure. The following proposition justifies in precise terms this way of sampling. Its proof is just an extension of the proof of Proposition 9.5.

Proposition 9.6 (GPS10)   *Fix a Boolean function f on Ω_n. For $A \subseteq \{1, 2, \ldots, n\}$ and $y \in \{\pm 1\}^{A^c}$, that is, a configuration on A^c, let g_y be the function defined on $\{\pm 1\}^A$ obtained by using f but fixing the configuration to be y outside of A. Then for any $S \subseteq A$, we have*

$$\hat{Q}(\mathscr{S}_f \cap A = S) = \mathbb{E}[\hat{Q}(\mathscr{S}_{g_y} = S)] = \mathbb{E}[\hat{g}_y^2(S)].$$

Proof   Using the first line of the proof of Proposition 9.5, it is easy to check that for any $S \subseteq A$, we have that

$$\mathbb{E}[f\chi_S \mid \mathcal{F}_{A^c}] = \sum_{S' \subseteq A^c} \widehat{f}(S \cup S')\chi_{S'}.$$

This gives

$$\mathbb{E}\Big[\mathbb{E}[f\chi_S \mid \mathcal{F}_{A^c}]^2\Big] = \sum_{S' \subseteq A^c} \widehat{f}(S \cup S')^2 = \hat{Q}[\mathscr{S} \cap A = S]$$

which is precisely the claim.   □

Remark   Observe that Proposition 9.5 is a special case of Proposition 9.6 when S is taken to be $\emptyset$ and A is replaced by A^c.

The following corollary was first observed by Gil Kalai.

Corollary 9.7 (GPS10)   *If f is a Boolean function mapping into $\{\pm 1\}$, then for all i and j,*

$$\mathbb{P}(i, j \in \mathcal{P}) = \hat{Q}(i, j \in \mathscr{S}).$$

(The comment immediately following Proposition 9.4 holds here as well.)

Proof   Although it has already been established that $\mathcal{P}$ and $\mathscr{S}$ have the same one-dimensional marginals, we first show how Proposition 9.6 can be used to establish this. This latter proposition yields, with $A = S = \{i\}$, that

$$\hat{Q}(i \in \mathscr{S}) = \hat{Q}(\mathscr{S} \cap \{i\} = \{i\}) = \mathbb{E}[\hat{g}_y^2(\{i\})].$$

Note that g_y is $\pm \omega_i$ if i is pivotal and constant if i is not pivotal. Hence the last term is $\mathbb{P}(i \in \mathcal{P})$.

For the two-dimensional marginals, one first checks this by hand when $n = 2$. For general n, taking $A = S = \{i, j\}$ in Proposition 9.6, we have

$$\hat{Q}(i, j \in \mathscr{S}) = \mathbb{P}(\mathscr{S} \cap \{i, j\} = \{i, j\}) = \mathbb{E}[\hat{g}_y^2(\{i, j\})].$$

For fixed y, the $n = 2$ case tells us that $\hat{g}_y^2(\{i, j\}) = \mathbb{P}(i, j \in \mathcal{P}_{g_y})$. Finally, a little thought shows that $\mathbb{E}[\mathbb{P}(i, j \in \mathcal{P}_{g_y})] = \mathbb{P}(i, j \in \mathcal{P})$, completing the proof.   □

9.3 Nontrivial spectrum near the upper bound for percolation

We now return to our central event of percolation crossings of the rectangle R_n where f_n denotes this event. At this point, we know that for $\mathbb{Z}^2$ (most of) the spectrum lies between n^{ϵ_0} (for some $\epsilon_0 > 0$) and $n^2 \alpha_4(n)$ while for $\mathbb{T}$ it sits between $n^{1/8 + o(1)}$ and $n^{3/4 + o(1)}$. In this section, we show that there is a nontrivial amount of spectrum near the upper bound $n^2 \alpha_4(n)$. For $\mathbb{T}$, in terms of quantitative noise sensitivity, this tells us that if our noise sequence ϵ_n is equal to $1/n^{3/4 - \delta}$ for fixed $\delta > 0$, then in the limit, the two variables $f(\omega)$ and $f(\omega_{\epsilon_n})$ are not perfectly correlated; that is, there is some degree of independence. (See the exercises for understanding such arguments.) However, we cannot conclude that there is full independence because we don't know that "all" of the spectrum is near $n^{3/4 + o(1)}$ (yet!).

Theorem 9.8 (GPS10)   *Consider our percolation crossing functions $\{f_n\}$ (with values into $\{\pm 1\}$) of the rectangles R_n for $\mathbb{Z}^2$ or $\mathbb{T}$. There exists $c > 0$ such that for all n,*

$$\hat{\mathbb{P}}[|\mathscr{S}_n| \geq c n^2 \alpha_4(n)] \geq c.$$

The key lemma for proving this is the following second moment bound on the number of pivotals which we prove afterwards. It has a similar flavor to Exercise 6.6 in Chapter 6.

Lemma 9.9 (GPS10)   *Consider our percolation crossing functions $\{f_n\}$ above and let R'_n be the box concentric with R_n with half the radius. If $X_n = |\mathcal{P}_n \cap R'_n|$ is the cardinality of the set of pivotal points in R'_n, then there exists a constant C such that for all n we have that*

$$\mathbb{E}[|X_n|^2] \leq C \mathbb{E}[|X_n|]^2.$$

Proof of Theorem 9.8. Since $\mathcal{P}_n$ and $\mathscr{S}_n$ have the same one and two-dimensional marginals, it follows fairly straightforward from Lemma 9.9 that we also have that for all n

$$\hat{\mathbb{P}}[|\mathscr{S}_n \cap R'_n|^2] \leq C \hat{\mathbb{P}}[|\mathscr{S}_n \cap R'_n|]^2.$$

Recall now the Paley–Zygmund inequality, which states that if $Z \geq 0$, then for all $\theta \in (0,1)$,

$$\mathbb{P}(Z \geq \theta \mathbb{E}[Z]) \geq (1-\theta)^2 \frac{\mathbb{E}[Z]^2}{\mathbb{E}[Z^2]}. \tag{9.3}$$

The two above inequalities (with $Z = |\mathscr{S}_n \cap R'_n|$ and $\theta = 1/2$) imply that for all n,

$$\hat{\mathbb{P}}[|\mathscr{S}_n \cap R'_n| \geq \frac{\hat{\mathbb{E}}[|\mathscr{S}_n \cap R'_n|]}{2}] \geq \frac{1}{4C}.$$

Now, by Proposition 9.4, one has that $\hat{\mathbb{E}}[|\mathscr{S}_n \cap R'_n|] = \mathbb{E}[X_n]$. Furthermore (a trivial modification of) Proposition 6.8 yields $\mathbb{E}[X_n] \asymp n^2 \alpha_4(n)$, which thus completes the proof. $\square$

We are now left with

Proof of Lemma 9.9. As indicated at the end of the proof of Theorem 9.8, we have that $\mathbb{E}(X_n) \asymp n^2 \alpha_4(n)$. Next, for $x, y \in R'_n$, a picture shows that

$$\mathbb{P}(x, y \in \mathcal{P}_n) \leq \alpha_4^2(|x-y|/2)\alpha_4(2|x-y|, n/2)$$

because we need to have the four-arm event around x to distance $|x-y|/2$, the same for y, and the four-arm event in the annulus centered at $(x+y)/2$ from distance $2|x-y|$ to distance $n/2$ and finally these three events are independent. This is by quasi-multiplicity at most

$$O(1)\alpha_4^2(n)/\alpha_4(|x-y|, n)$$

and hence

$$\mathbb{E}[|X_n|^2] \leq O(1)\alpha_4^2(n) \sum_{x,y} \frac{1}{\alpha_4(|x-y|, n)}.$$

Since, for a given x, there are at most $O(1)2^{2k}$ y's with $|x-y| \in [2^k, 2^{k+1}]$, using quasi-multiplicity, the above sum is at most

$$O(1)n^2\alpha_4^2(n) \sum_{k=0}^{\log_2(n)} \frac{2^{2k}}{\alpha_4(2^k, n)}.$$

Using

$$\frac{1}{\alpha_4(r,R)} \leq (R/r)^{2-\epsilon}$$

(this is the fact that the four-arm exponent is strictly less than 2), the sum becomes at most

$$O(1)n^{4-\epsilon}\alpha_4^2(n) \sum_{k=0}^{\log_2(n)} 2^{k\epsilon}.$$

Since the last sum is at most $O(1)n^\epsilon$, we are done.   □

In terms of the consequences for quantitative noise sensitivity, Theorem 9.8 implies the following corollary; see the exercises for similar implications.

Corollary 9.10   *For $\mathbb{T}$ or $\mathbb{Z}^2$, there exists $c > 0$ so that if $\epsilon_n = 1/(n^2\alpha_4(n))$, then for all n,*

$$\mathbb{P}(f_n(\omega) \neq f_n(\omega_{\epsilon_n})) \geq c.$$

Note, importantly, this does *not* say that $f_n(\omega)$ and $f_n(\omega_{\epsilon_n})$ become asymptotically uncorrelated, only that they are not asymptotically completely correlated. To ensure that they are asymptotically uncorrelated is significantly more difficult and requires showing that "all" of the spectrum is near $n^{3/4}$. This much more difficult task is the subject of the next chapter.

9.4 Two more facts concerning the spectral sample in general

The two following propositions were used in (GPS10) for studying percolation crossings; because these potentially could be used in more general situations, we record the following results here.

Proposition 9.11 (GPS10)   *Let f be a monotone Boolean function mapping Ω_n into $\{\pm 1\}$. Let W be a fixed subset of $[n]$ and let (ω_1, ω_2) be a coupling of two i.i.d. percolations on $[n]$ which are such that*

$$\begin{cases} \omega_1 = \omega_2 & \text{on } W^c \\ \omega_1, \omega_2 & \text{are independent on } W \end{cases}$$

Then for $x \notin W$:

$$\hat{\mathbb{Q}}[x \in \mathscr{S}_f \text{ and } \mathscr{S}_f \cap W = \emptyset] = \mathbb{P}[x \text{ is pivotal for } \omega_1 \text{ and } \omega_2].$$

Proof   This follows fairly easily from Proposition 9.6.   □
Remark This identity in the special cases where $W = \emptyset$ or $W = \{x\}^c$ we have seen much earlier.

Let f be a Boolean function mapping Ω into $\{\pm 1\}$. For B a subset of the bits, define $\Lambda_B = \Lambda_{f,B}$ as the event that B is pivotal for f. More precisely, Λ_B is the set of $\omega \in \Omega$ such that there is some $\omega' \in \Omega$ that agrees with ω on B^c while $f(\omega) \neq f(\omega')$. Let W also be a subset of the bits and and let (ω_1, ω_2) be as in Proposition 9.11. Finally, assuming W and B are disjoint, let $g(B, W) := \mathbb{P}[B \text{ is pivotal for } \omega_1 \text{ and } \omega_2]$.

Proposition 9.12 (GPS10) *Let $\mathscr{S} = \mathscr{S}_f$ be the spectral sample of some $f: \Omega \to \mathbb{R}$, and let W and B be disjoint subsets of the bits. Then*

$$\hat{\mathbb{Q}}[\mathscr{S} \cap B \neq \emptyset = \mathscr{S} \cap W] \leq 4 \|f\|_\infty^2 \, g(B, W).$$

Proof Proposition 9.5 first yields

$$\hat{\mathbb{Q}}[\mathscr{S} \cap B \neq \emptyset, \ \mathscr{S} \cap W = \emptyset] = \hat{\mathbb{Q}}[\mathscr{S} \subseteq W^c] - \hat{\mathbb{Q}}[\mathscr{S} \subseteq (W \cup B)^c]$$

$$= \mathbb{E}\Big[\mathbb{E}[f \mid \mathcal{F}_{W^c}]^2 - \mathbb{E}[f \mid \mathcal{F}_{(W \cup B)^c}]^2\Big] \qquad (9.4)$$

$$= \mathbb{E}\Big[\big(\mathbb{E}[f \mid \mathcal{F}_{W^c}] - \mathbb{E}[f \mid \mathcal{F}_{(W \cup B)^c}]\big)^2\Big].$$

On the complement of Λ_B, we have $f = \mathbb{E}[f \mid \mathcal{F}_{B^c}]$. Therefore,

$$-2\|f\|_\infty 1_{\Lambda_B} \leq f - \mathbb{E}[f \mid \mathcal{F}_{B^c}] \leq 2\|f\|_\infty 1_{\Lambda_B}.$$

Taking conditional expectations throughout, we get

$$-2\|f\|_\infty \mathbb{P}[\Lambda_B \mid \mathcal{F}_{W^c}] \leq \mathbb{E}[f \mid \mathcal{F}_{W^c}] - \mathbb{E}[\mathbb{E}[f \mid \mathcal{F}_{B^c}] \mid \mathcal{F}_{W^c}] \leq 2\|f\|_\infty \mathbb{P}[\Lambda_B \mid \mathcal{F}_{W^c}].$$

Note that $\mathbb{E}[\mathbb{E}[f \mid \mathcal{F}_{B^c}] \mid \mathcal{F}_{W^c}] = \mathbb{E}[f \mid \mathcal{F}_{(B \cup W)^c}]$, since our measure on Ω is i.i.d. Thus, the above gives

$$\Big|\mathbb{E}[f \mid \mathcal{F}_{W^c}] - \mathbb{E}[f \mid \mathcal{F}_{(B \cup W)^c}]\Big| \leq 2\|f\|_\infty \mathbb{P}[\Lambda_B \mid \mathcal{F}_{W^c}].$$

An appeal to (9.4) now completes the proof, by observing that $g(B, W)$ is precisely the second moment of $\mathbb{P}[\Lambda_B \mid \mathcal{F}_{W^c}]$. □

9.5 Exercises

9.1 Let $\{f_n\}$ be an arbitrary sequence of Boolean functions mapping into $\{\pm 1\}$ with corresponding spectral samples $\{\mathscr{S}_n\}$.
(a) Show that $\hat{\mathbb{P}}[0 < |\mathscr{S}_n| \leq A_n] \to 0$ implies that $\hat{\mathbb{E}}[(1 - \epsilon_n)^{|\mathscr{S}_n|} I_{\mathscr{S}_n \neq \emptyset}] \to 0$ if $\epsilon_n A_n \to \infty$.
(b) Show that $\hat{\mathbb{E}}[(1 - \epsilon_n)^{|\mathscr{S}_n|} I_{\mathscr{S}_n \neq \emptyset}] \to 0$ implies that $\hat{\mathbb{P}}[0 < |\mathscr{S}_n| \leq A_n] \to 0$ if $\epsilon_n A_n \leq O(1)$.

9.2 Let $\{f_n\}$ be an arbitrary sequence of Boolean functions mapping into $\{\pm 1\}$ with corresponding spectral samples $\{\mathscr{S}_n\}$.
(a) Show that $\mathbb{P}[f(\omega) \neq f(\omega_{\epsilon_n})] \to 0$ and $A_n \epsilon_n \geq \Omega(1)$ imply that

$$\hat{\mathbb{P}}[|\mathscr{S}_n| \geq A_n] \to 0.$$

(b) Show that $\hat{\mathbb{P}}[|\mathscr{S}_n| \geq A_n] \to 0$ and $A_n \epsilon_n = o(1)$ imply that

$$\mathbb{P}[f(\omega) \neq f(\omega_{\epsilon_n})] \to 0.$$

9.3 Prove Corollary 9.10.

9.4 For the Iterated 3-Majority sequence, recall that the total influence is n^α where $\alpha = 1 - \log 2/\log 3$. Show that for $\epsilon_n = 1/n^\alpha$, $\mathbb{P}(f_n(\omega) \neq f_n(\omega_{\epsilon_n}))$ does not tend to 0.

9.5 Assume that $\{f_n\}$ is a sequence of monotone Boolean functions on n bits with total influence equal to $n^{1/2}$ up to constants. Show that the sequence cannot be noise sensitive. Is it necessarily noise stable as the Majority function is?

9.6 Assume that $\{f_n\}$ is a sequence of monotone Boolean functions with mean 0 on n bits. Show that one cannot have noise sensitivity when using noise level $\epsilon_n = 1/n^{1/2}$.

9.7 Show that $\mathcal{P}$ and $\mathcal{S}$ have the same two-dimensional marginals using only Proposition 9.5 rather than Proposition 9.6.
Hint: It suffices to show that $\mathbb{P}(\{i,j\} \cap \mathcal{P} = \emptyset) = \hat{\mathbb{Q}}(\{i,j\} \cap \mathcal{S} = \emptyset)$.

9.8 Fill in the details on how one obtains Proposition 9.11 from Proposition 9.6.

9.9 For the Iterated 3-Majority function with k levels, it turns out that the pivotal set and the spectrum have a very nice probabilistic description.
(a) Show that the distribution of the pivotal set is the same as the distribution of the kth generation of a Galton–Watson process with offspring distribution $\frac{1}{4}\delta_0 + \frac{3}{4}\delta_2$.
(b) Show that the distribution of the spectral sample is the same as the distribution of the kth generation of a Galton–Watson process with offspring distribution $\frac{3}{4}\delta_1 + \frac{1}{4}\delta_3$.

9.10 If f is a function with $E(f^2) < 1$ (e.g., if f is the indicator function of a nontrivial event), then we still have the spectral measure but this becomes a subprobability measure (of total weight $E(f^2)$) rather than a probability measure. For this reason, sometimes when dealing with events A, it can be convenient to deal with the function $I_A - I_{A^c}$ (which is the function that is 1 on A and -1 on A^c) because this function is always ± 1 and hence its spectrum is a probability measure. Describe the exact relationship between the spectral (sub)probability measure corresponding to I_A and the spectral probability measure corresponding to $I_A - I_{A^c}$.

9.11 (Challenging problem) Do you expect that Exercise 9.5 is sharp, meaning that, if $1/2$ is replaced by $\alpha < 1/2$, then one can find noise-sensitive examples?

10

Sharp noise sensitivity of percolation

We explain in this chapter the main ideas of the proof in (GPS10) that most of the "spectral mass" lies near $n^2\alpha_4(n) \approx n^{3/4+o(1)}$. As this proof is rather long and involved, the content of this chapter is far from a formal proof. Rather it should be considered as a (hopefully convincing) heuristic explanation of the main results, and possibly for the interested readers as a "reading guide" for the paper (GPS10).

Very briefly speaking, the idea behind the proof is to identify properties of the geometry of $\mathscr{S}_{f_n}$ which are reminiscent of a self-similar fractal structure. Ideally, $\mathscr{S}_{f_n}$ would behave like a spatial branching tree (or in other words a fractal percolation process), where distinct branches evolve independently of each other. This is conjecturally the case, but it turns out that it is very hard to control the dependency structure within $\mathscr{S}_{f_n}$. In (GPS10), only a tiny hint of spatial *independence* within $\mathscr{S}_{f_n}$ is proved. One of the main difficulties of the proof is to overcome the fact that one has very little independence to play with.

A substantial part of this chapter focuses on the much simpler case of fractal percolation. Indeed, this process can be seen as the simplest toy model for the spectral sample $\mathscr{S}_{f_n}$. Explaining the simplified proof adapted to this setting already enables us to convey some of the main ideas for handling $\mathscr{S}_{f_n}$.

10.1 State of the art and main statement

See Figure 10.1 where we summarize what we have learned so far about the spectral sample $\mathscr{S}_{f_n}$ of a left to right crossing event f_n.

From this table, we see that the main question now is to prove that all the spectral mass indeed diverges at speed $n^2\alpha_4(n)$ which is $n^{3/4+o(1)}$ for the triangular lattice. This is the content of the following theorem.

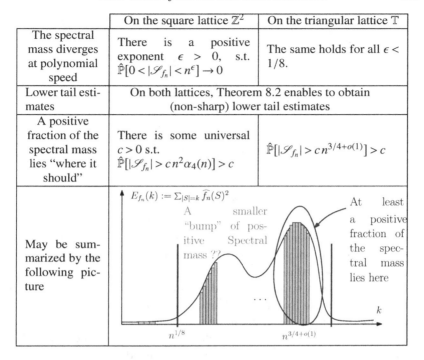

	On the square lattice $\mathbb{Z}^2$	On the triangular lattice $\mathbb{T}$
The spectral mass diverges at polynomial speed	There is a positive exponent $\epsilon > 0$, s.t. $\hat{\mathbb{P}}[0 < |\mathscr{S}_{f_n}| < n^\epsilon] \to 0$	The same holds for all $\epsilon < 1/8$.
Lower tail estimates	On both lattices, Theorem 8.2 enables to obtain (non-sharp) lower tail estimates	
A positive fraction of the spectral mass lies "where it should"	There is some universal $c > 0$ s.t. $\hat{\mathbb{P}}[|\mathscr{S}_{f_n}| > cn^2\alpha_4(n)] > c$	$\hat{\mathbb{P}}[|\mathscr{S}_{f_n}| > cn^{3/4+o(1)}] > c$
May be summarized by the following picture		

Figure 10.1 A summary of some of the results obtained so far for $\mathscr{S}_{f_n}$.

Theorem 10.1 (GPS10)

$$\limsup_{n \to \infty} \hat{\mathbb{P}}[0 < |\mathscr{S}_{f_n}| < \lambda n^2\alpha_4(n)] \xrightarrow[\lambda \to 0]{} 0 .$$

On the triangular lattice $\mathbb{T}$, the rate of decay in λ is known explicitly. Namely:

Theorem 10.2 (GPS10) *On the triangular grid $\mathbb{T}$, the lower tail of $|\mathscr{S}_{f_n}|$ satisfies*

$$\limsup_{n \to \infty} \hat{\mathbb{P}}[0 < |\mathscr{S}_{f_n}| < \lambda \hat{\mathbb{E}}[|\mathscr{S}_{f_n}|])] \underset{\lambda \to 0}{\asymp} \lambda^{2/3} .$$

This result deals with what one might call the "macroscopic" lower tail, that is, with quantities which asymptotically are still of order $\hat{\mathbb{E}}[|\mathscr{S}_{f_n}|]$ (since λ remains fixed in the above lim sup). It turns out that in our later study of dynamical percolation in Chapter 11, we will need a sharp control on the full lower tail. This is the content of the following stronger theorem:

Theorem 10.3 (GPS10)   *On $\mathbb{Z}^2$ and on the triangular grid $\mathbb{T}$, for all $1 \leq r \leq n$, one has*

$$\hat{\mathbb{P}}[0 < |\mathscr{S}_{f_n}| < r^2 \alpha_4(r)] \asymp \frac{n^2}{r^2} \alpha_4(r,n)^2.$$

On the triangular grid, this translates into

$$\hat{\mathbb{P}}[0 < |\mathscr{S}_{f_n}| < u] \approx n^{-\frac{1}{2}} u^{\frac{2}{3}},$$

where we write $\approx$ to avoid relying on $o(1)$ terms in the exponents.

10.2 Overall strategy

In the above theorems, it is clear that we are mostly interested in the cardinality of $\mathscr{S}_{f_n}$. However, our strategy will consist in understanding as much as we can about the typical *geometry* of the random set $\mathscr{S}_{f_n}$ sampled according to the spectral probability measure $\hat{\mathbb{P}}_{f_n}$.

As we have seen so far, the random set $\mathscr{S}_{f_n}$ shares many properties with the set of pivotal points $\mathcal{P}_{f_n}$. A first possibility would be that they are asymptotically similar. After all, noise sensitivity is intimately related with pivotal points, so it is not unreasonable to hope for such a behavior. This scenario would be very convenient for us because the geometry of $\mathcal{P}_{f_n}$ is now well understood (at least on $\mathbb{T}$) thanks to the SLE processes. In particular, in the case of $\mathcal{P}_{f_n}$, one can "explore" $\mathcal{P}_{f_n}$ in a Markovian way by relying on exploration processes. Unfortunately, based on very convincing heuristics, it is conjectured that the scaling limits of $\frac{1}{n}\mathscr{S}_{f_n}$ and $\frac{1}{n}\mathcal{P}_{f_n}$ are singular random compact sets of the square. See Figure 10.2 for a quick overview of the similarities and differences between these two random sets.

The conclusion of this table is that they indeed share many properties, but one cannot deduce lower tail estimates on $|\mathscr{S}_{f_n}|$ out of lower tail estimates on $|\mathcal{P}_{f_n}|$. Also, even worse, we will not be allowed to rely on spatial Markov properties for $\mathscr{S}_{f_n}$.

However, even though $\mathcal{P}_{f_n}$ and $\mathscr{S}_{f_n}$ differ in many ways, they share at least one essential property: a seemingly *self-similar fractal behavior*. The main strategy in (GPS10) to control the lower-tail behavior of $|\mathscr{S}_{f_n}|$ is to prove that *in some very weak sense*, $\mathscr{S}_{f_n}$ behaves like the simplest model among self-similar fractal processes in $[0,n]^2$: that is, a supercritical spatial Galton–Watson tree embedded in $[0,n]^2$, also called a *fractal percolation process*. The lower tail of this very simple toy model is investigated in detail in the next section with a technique that will be suitable for $\mathscr{S}_{f_n}$. The main difficulty that arises in this program is the lack of knowledge

	Pivotal set $\mathcal{P}_{fn}$	Spectral set $\mathcal{S}_{fn}$
First moment	$\mathbb{E}\left[|\mathcal{P}_{fn}|\right]$	$\hat{\mathbb{E}}\left[|\mathcal{S}_{fn}|\right]$
Second moment	$\mathbb{E}\left[|\mathcal{P}_{fn}|^2\right]$	$\hat{\mathbb{E}}\left[|\mathcal{S}_{fn}|^2\right]$
Higher moments $(k \geq 3)$	In general, they differ!	
Methods for sampling these random sets	Easy (and fast) using two exploration paths:	The spectral sample $\mathcal{S}_{fn}$ is much harder to sample. In fact, the only known way to proceed is to compute the weights $\hat{f}_n(S)^2$, one at a time ...
Spatial correlation structure	Distant regions in $\mathcal{P}_{fn}$ behave more or less independently of each other. Furthermore, one can use the very convenient spatial Markov property due to the i.i.d structure of the percolation picture.	? Much of the picture here remains unclear.
Lower tail behavior	$\mathbb{P}\left[|\mathcal{P}_{fn}|=1\right] \asymp n^{-\frac{11}{12}}$	$\hat{\mathbb{P}}\left[|\mathcal{S}_{fn}|=1\right] \asymp n^{-\frac{1}{2}}$

Figure 10.2 Similarities and differences between $\mathcal{S}_{fn}$ and $\mathcal{P}_{fn}$.

of the independency structure within $\mathcal{S}_{f_n}$. In other words, when we try to compare $\mathcal{S}_{f_n}$ with a fractal percolation process, the self-similarity already requires some work, but the hardest part is to deal with the fact that distinct "branches" (or rather their analogs) are not known to behave even slightly independently of each other. We discuss these issues in Section 10.4 but do not give a complete proof.

10.3 Toy model: The case of fractal percolation

As we explained previously, our main strategy is to exploit the fact that $\mathcal{S}_{f_n}$ has a certain self-similar fractal structure. Along this section, we consider the simplest case of such a self-similar fractal object: namely *fractal percolation*, and we detail in this simple setting what our later strategy will be. Deliberately, this strategy will not be optimal in this simplified case. In particular, we do not rely on the martingale techniques that one can use with fractal percolation or Galton–Watson trees, as such methods would not be available for our spectral sample $\mathscr{S}_{f_n}$. Results of this type can be found in (FW07) or in (RW10) for the Poisson case.

10.3.1 Definition of the model and first properties

To make the analogy with $\mathscr{S}_{f_n}$ easier let

$$n := 2^h, h \geq 1,$$

and let's fix a parameter $p \in (0, 1)$.

Now, *fractal percolation* on $[0, n]^2$ is defined inductively as follows: divide $[0, 2^h]^2$ into 4 squares and retain each of them independently with probability p. Let $\mathcal{T}^1$ be the union of the retained 2^{h-1}-squares. The second-level tree $\mathcal{T}^2$ is obtained by reiterating the same procedure independently for each 2^{h-1}-square in $\mathcal{T}^1$. Continuing in the same fashion all the way to the squares of unit size, one obtains $\mathcal{T}_n = \mathcal{T} := \mathcal{T}^h$ which is a random subset of $[0, n]^2$. See (Lyo11) for more on the definition of *fractal percolation*. See also Figure 10.3 for an example of $\mathcal{T}^5$.

Remark 10.4   We thus introduced two different notations for the same random set ($\mathcal{T}_{n=2^h} \equiv \mathcal{T}^h$). The reason for this is that on the one hand the notation $\mathcal{T}_n$ defined on $[0, n]^2 = [0, 2^h]^2$ makes the analogy with $\mathscr{S}_{f_n}$ (also defined on $[0, n]^2$) easier, while on the other hand inductive proofs will be more convenient with the notation $\mathcal{T}^h$.

To have a supercritical Galton–Watson tree, one has to choose $p \in (1/4, 1)$. Furthermore, one can easily check the following easy proposition.

Proposition 10.5   *Let* $p \in (1/4, 1)$. *Then*

$$\mathbb{E}[|\mathcal{T}_n|] = n^2 p^h = n^{2 + \log_2 p},$$

and

$$\mathbb{E}[|\mathcal{T}_n|^2] \leq O(1) \mathbb{E}[|\mathcal{T}_n|]^2.$$

In particular, by the second moment method (e.g., the Paley–Zygmund inequality), with positive probability, $\mathcal{T}_n$ *is of order* $n^{2 + \log_2 p}$.

Let

$$\alpha := 2 + \log_2 p.$$

This parameter α corresponds to the "fractal dimension" of $\mathcal{T}_n$. To make the analogy with $\mathscr{S}_{f_n}$ even clearer, one could choose p in such a way that $\alpha = 2 + \log_2 p = 3/4$, but we will not need to.

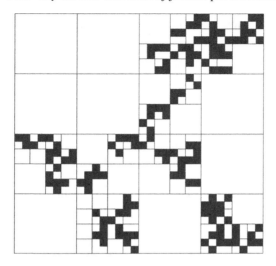

Figure 10.3 A realization of a fractal percolation $\mathcal{T}_{2^5} = \mathcal{T}^5$.

The above proposition implies that on the event $\mathcal{T}_n \neq \emptyset$, with positive conditional probability $|\mathcal{T}_n|$ is large (of order n^α). This is the exact analog of Theorem 9.8 for the spectral sample $\mathscr{S}_{f_n}$.

Let us first analyze what would be the analog of Theorem 10.1 in the case of our toy model $\mathcal{T}_n$. We have the following.

Proposition 10.6

$$\limsup_{n\to\infty} \mathbb{P}[0 < |\mathcal{T}_n| < \lambda n^\alpha)] \underset{\lambda\to 0}{\longrightarrow} 0 .$$

Remark If one could rely on martingale techniques, then this proposition is a corollary of standard results. Indeed, as is well-known

$$M_i := \frac{|\mathcal{T}^i|}{(4p)^i} ,$$

is a positive martingale. Therefore it converges, as $n \to \infty$, to a nonnegative random variable $W \geq 0$. Furthermore, the conditions of the Kesten–Stigum Theorem are fulfilled (see, e.g., Section 5.1 in (Lyo11)) and therefore W is positive on the event that there is no extinction. This implies the above proposition.

As we claimed previously, we will intentionally follow a more hands-on approach in this section that will be more suitable to the random set

How does $\mathcal{L}(\mathcal{T}_n \mid 0 < |\mathcal{T}_n| < u)$ look ?

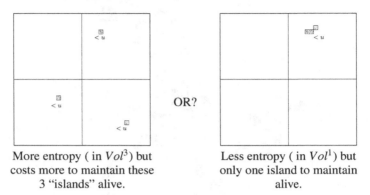

More entropy (in Vol^3) but
costs more to maintain these
3 "islands" alive.

Less entropy (in Vol^1) but
only one island to maintain
alive.

Figure 10.4 Entropy vs. clustering effect.

$\mathscr{S}_{f_n}$ that we have in mind. Furthermore, this approach will have the great advantage to provide the following much more precise result, which is the analog of Theorem 10.3 for $\mathcal{T}_n$.

Proposition 10.7 *For any* $1 \le r \le n,$

$$\mathbb{P}[0 < |\mathcal{T}_n| < r^\alpha] \asymp (\frac{r}{n})^{\log_2 1/\mu},$$

where μ *is an explicit constant in* $(0, 1)$ *computed in Exercise 10.2.*

10.3.2 Strategy and heuristics

Letting $u \ll n^\alpha$, we wish to estimate $\mathbb{P}[0 < |\mathcal{T}_n| < u]$. Even though we are interested only in the size of $\mathcal{T}_n$, we will try to estimate this quantity by understanding the *geometry* of the conditional set:

$$\mathcal{T}_n^{|u} := \mathcal{L}(\mathcal{T}_n \mid 0 < |\mathcal{T}_n| < u).$$

The first natural question to ask is whether this conditional random set is typically *localized* or not. See Figure 10.4.

Intuitively, it is quite clear that the set $\mathcal{T}_n$ conditioned to be very small will tend to be localized. So it is the picture on the right in Figure 10.4 which is more likely. This would deserve a proof of course, but we come back to this later. The fact that it should look more and more localized tells us that as one shrinks u, this should make our conditional $\mathcal{T}_n^{|u}$ more and more singular with respect to the unconditional one. But how much localization should we see? This is again fairly easy to answer, at least on the intuitive level. Indeed, $\mathcal{T}_n^{|u}$ should tend to localize until it reaches a certain mesoscopic scale r such that $1 \ll r \ll n$. One can compute how much it costs to maintain a single branch (or $O(1)$ branches) alive until scale r, but once this is achieved, one should let the system evolve in a "natural" way. In particular, once the tree survives all the way to a mesoscopic square of size r, it will (by the second moment method) produce $\Omega(r^\alpha)$ leaves there with positive probability.

To summarize, typically $\mathcal{T}_n^{|u}$ will maintain $O(1)$ many branches alive at scale $1 \ll r \ll n$, and then it will let the branching structure evolve in a basically unconditional way. The intermediate scale r is chosen so that $r^\alpha \asymp u$.

Definition 10.8   If $1 \leq r \leq n = 2^h$ is such that $r = 2^l, 0 \leq l \leq h$, let $\mathcal{T}_{(r)}$ denote the set of branches that were still alive at scale $r = 2^l$ in the iterative construction of $\mathcal{T}_n$. In other words, $\mathcal{T}_{(r)} \equiv \mathcal{T}^{h-l}$ and $\mathcal{T}_n \subset \bigcup \mathcal{T}_{(r)}$. This random set $\mathcal{T}_{(r)}$ will be the analog of the "r-smoothing" $\mathcal{S}_{(r)}$ of the spectral sample $\mathcal{S}_{f_n}$ defined later in Definition 10.13.

Returning to our problem, the above heuristics say that one expects to have for any $1 \ll u \ll n^\alpha$.

$$\mathbb{P}[0 < |\mathcal{T}_n| < u] \asymp \mathbb{P}[0 < |\mathcal{T}_{(r)}| \leq O(1)]$$
$$\asymp \mathbb{P}[|\mathcal{T}_{(r)}| = 1],$$

where r is a dyadic integer chosen such that $r^\alpha \asymp u$. Or in other words, we expect that

$$\mathbb{P}[0 < |\mathcal{T}_n| < r^\alpha] \asymp \mathbb{P}[|\mathcal{T}_{(r)}| = 1]. \tag{10.1}$$

In the next subsection, we briefly explain how this heuristic can be implemented into a proof in the case of the tree $\mathcal{T}_n$ in a way which will be suitable to the study of $\mathcal{S}_{f_n}$. We only skim through the main ideas for this tree case.

10.3.3 Setup of a proof for $\mathcal{T}_n$

Motivated by the preceding heuristics, we divide our system into two scales: above and below the mesoscopic scale r. One can write the lower tail event as follows (let $1 \ll r \ll n$):

$$\mathbb{P}[0 < |\mathcal{T}_n| < r^{\alpha}] = \sum_{k \geq 1} \mathbb{P}[|\mathcal{T}_{(r)}| = k]\mathbb{P}[0 < |\mathcal{T}_n| < r^{\alpha} \,|\, |\mathcal{T}_{(r)}| = k]. \qquad (10.2)$$

It is not hard to estimate the second term $\mathbb{P}[0 < |\mathcal{T}_n| < r^{\alpha} \,|\, |\mathcal{T}_{(r)}| = k]$. Indeed, in this term we are conditioning on having exactly k branches alive at scale r. Independently of where they are, "below" r, these k branches evolve independently of each other. Furthermore, by the second moment method, there is a universal constant $c > 0$ such that each of them exceeds the fatal amount of r^{α} leaves with probability at least c (note that in the opposite direction, each branch could also go extinct with positive probability). This implies that

$$\mathbb{P}[0 < |\mathcal{T}_n| < r^{\alpha} \,|\, |\mathcal{T}_{(r)}| = k] \leq (1 - c)^k.$$

Remark Note that one makes heavy use of the independence structure within $\mathcal{T}_n$ here. This aspect is much more nontrivial for the spectral sample $\mathscr{S}_{f_n}$. Fortunately it turns out, and this is a key fact, that in (GPS10) one can prove a weak independence statement that in some sense makes it possible to follow this route.

We are left with the following upper bound:

$$\mathbb{P}[0 < |\mathcal{T}_n| < r^{\alpha}] \leq \sum_{k \geq 1} \mathbb{P}[|\mathcal{T}_{(r)}| = k](1 - c)^k. \qquad (10.3)$$

To prove our goal of (10.1), by exploiting the exponential decay given by $(1 - c)^k$ (which followed from independence), it is enough to prove the following bound on the mesoscopic behavior of $\mathcal{T}$:

Lemma 10.9 *There is a **subexponential** function $k \mapsto g(k)$ such that for all $1 \leq r \leq n$,*

$$\mathbb{P}[|\mathcal{T}_{(r)}| = k] \leq g(k)\mathbb{P}[|\mathcal{T}_{(r)}| = 1].$$

Notice as we did in Definition 10.8 that since $\mathcal{T}_{(r)}$ has the same law as $\mathcal{T}^{h-l}$, this is a purely Galton–Watson tree type of question.

The big advantage of our strategy so far is that initially we were looking for a sharp control on $\mathbb{P}[0 < |\mathcal{T}_n| < u]$ and now, using this "two-scales" argument, it only remains to prove a crude upper bound on the lower tail of $|\mathcal{T}_{(r)}|$. By scale invariance this is nothing else than obtaining a crude upper bound on the lower tail of $|\mathcal{T}_n|$. Hence this division into two scales greatly simplified our task.

10.3.4 Subexponential estimate on the lower-tail (Lemma 10.9)

The first step toward proving and understanding Lemma 10.9 is to understand the term $\mathbb{P}[|\mathcal{T}_{(r)}| = 1]$. From now on, it will be easier to work with the "dyadic" notations instead, that is, with $\mathcal{T}^i \equiv \mathcal{T}_{2^i}$ (see Remark 10.4). With these notations, the first step is equivalent to understanding the probabilities $p_i := \mathbb{P}[|\mathcal{T}^i| = 1]$. This aspect of the problem is very specific to the case of Galton–Watson trees and gives very little insight into the later study of the spectrum $\mathscr{S}_{f_n}$. Therefore we postpone the details to Exercise 10.2. The conclusion of this (straightforward) exercise is that p_i behaves as $i \to \infty$ like

$$p_i \sim c\mu^i,$$

for an explicit exponent $\mu \in (0, 1)$ (see Exercise 10.2). In particular, in order to prove Proposition 10.7, it is now enough to find a subexponential function $k \mapsto g(k)$ such that for any $i, k \geq 1$,

$$\mathbb{P}[|\mathcal{T}^i| = k] \leq g(k)\mu^i. \tag{10.4}$$

More precisely, we prove the following lemma.

Lemma 10.10 *Let* $g(k) := 2^{\theta \log_2^2(k+2)}$, *where* θ *is a fixed constant to be chosen later. Then for all* $i, k \geq 1$, *one has*

$$\mathbb{P}[|\mathcal{T}^i| = k] \leq g(k)\mu^i. \tag{10.5}$$

We provide the proof of this lemma here, as it can be seen as a "toy proof" of the corresponding subexponential estimate needed for the r-smoothed spectral samples $\mathscr{S}_{(r)}$, stated in the coming Theorem 10.16. The proof of this latter theorem shares some similarities with the proof below but is much more technical because in the case of $\mathscr{S}_{(r)}$ one has to deal with a more complex structure than the branching structure of a Galton–Watson tree.

Proof We proceed by double induction. Let $k \geq 2$ be fixed and assume that equation (10.5) is already satisfied for all pair (i', k') such that $k' < k$. Based on this assumption, let us prove by induction on i that all pairs (i, k) satisfy equation (10.5) as well.

First of all, if i is small enough, this is obvious by the definition of $g(k)$. Let

$$J = J_k := \sup\{i \geq 1 : g(k)\mu^i > 10\}.$$

Then, it is clear that equation (10.5) is satisfied for all (i,k) with $i \leq J_k$. Now let $i > J_k$.

If $\mathcal{T}^i$ is such that $|\mathcal{T}^i| = k \geq 1$, let $L = L(\mathcal{T}^i) \geq 0$ be the largest integer such that $\mathcal{T}^i$ intersects only one square of size 2^{i-L}. This means that below scale 2^{i-L}, the tree $\mathcal{T}^i$ splits into at least two live branches in distinct dyadic squares of size 2^{i-L-1}. Let $d \in \{2,3,4\}$ be the number of such live branches. By decomposing on the value of L, and using the above assumption, we get

$$\mathbb{P}[|\mathcal{T}^i| = k] \leq \mathbb{P}[L(\mathcal{T}^i) > i - J_k] +$$

$$\frac{1}{1-q} \sum_{l=0}^{i-J_k} \mathbb{P}[L(\mathcal{T}^i) = l] \sum_{d=2}^{4} \binom{4}{d} (\mu^{i-l-1})^d \sum_{\substack{(k_j)_{1 \leq j \leq d} \\ k_j \geq 1, \sum k_j = k}} \prod_j g(k_j)$$

where q is the probability that our Galton–Watson tree goes extinct.

Let us first estimate what $\mathbb{P}[L(\mathcal{T}^i) \geq m]$ is for $m \geq 0$. If $m \geq 1$, this means that among the 2^{2m} dyadic squares of size 2^{i-m}, only one will remain alive all the way to scale 1. Yet, it might be that some other such squares are still alive at scale 2^{i-m} but will go extinct by the time they reach scale 1. Let $p_{m,b}$ be the probability that the process $\mathcal{T}^{m+b}$ which lives in $[0, 2^{m+b}]^2$, is entirely contained in a dyadic square of size 2^b. With such notations, one has

$$\mathbb{P}[L(\mathcal{T}^i) \geq m] = p_{m,i-m}.$$

Furthermore, if $i = m$, one has $p_{i,0} = p_i \sim c\mu^i$. It is not hard to prove (see Exercise 10.2) the following lemma.

Lemma 10.11 *For any value of $m, b \geq 0$, one has*

$$p_{m,b} \leq \mu^m.$$

In particular, one has a universal upper bound in $b \geq 0$.

It follows from the lemma that $\mathbb{P}[L(\mathcal{T}^i) = l] \leq \mathbb{P}[L(\mathcal{T}^i) \geq l] \leq \mu^l$ and

$$\mathbb{P}[L(\mathcal{T}^i) > i - J_k] \leq \mu^{i-J_k} \tag{10.6}$$

$$\leq \frac{1}{10} g(k) \mu^i \text{ by the definition of } J_k. \tag{10.7}$$

This gives us that for some constant C

$$\mathbb{P}[|\mathcal{T}^i| = k] \leq \frac{\mu^i}{10} g(k) + C \sum_{l=0}^{i-J_k} \mu^l \sum_{d=2}^{4} (\mu^{i-l})^d \sum_{\substack{(k_j)_{1 \leq j \leq d} \\ k_j \geq 1, \sum k_j = k}} \prod_j g(k_j)$$

$$= \frac{\mu^i}{10} g(k) + C \mu^i \sum_{d=2}^{4} \sum_{l=0}^{i-J_k} (\mu^{i-l})^{d-1} \sum_{\substack{(k_j)_{1 \leq j \leq d} \\ k_j \geq 1, \sum k_j = k}} \prod_j g(k_j).$$

Let us deal with the $d = 2$ sum (the contributions coming from $d > 2$ being even smaller). By concavity of $k \mapsto \theta \log_2^2(k + 2)$, one obtains that for any (k_1, k_2) such that $k_1 + k_2 = k$: $g(k_1)g(k_2) \leq g(k/2)^2$. Because there are at most k^2 such pairs, this gives us the following bound on the $d = 2$ sum.

$$\sum_{l=0}^{i-J_k} (\mu^{i-l})^{2-1} \sum_{\substack{(k_j)_{1 \leq j \leq 2} \\ k_j \geq 1, \sum k_j = k}} \prod_j g(k_j) \leq \sum_{l=0}^{i-J_k} \mu^{i-l} k^2 g(k/2)^2$$

$$\leq \frac{1}{1-\mu} \mu^{J_k} k^2 g(k/2)^2$$

$$\leq 10 \frac{1}{1-\mu} k^2 g(k/2)^2 (\mu g(k))^{-1},$$

by definition of J_k.

Now, some easy analysis implies that if one chooses the constant $\theta > 0$ large enough, then for any $k \geq 2$, one has $C 10 \frac{1}{1-\mu} k^2 g(k/2)^2 (\mu g(k))^{-1} \leq \frac{1}{10} g(k)$. Altogether (and taking into consideration the $d > 2$ contributions), this implies that

$$\mathbb{P}[|\mathcal{T}^i| = k] \leq \frac{2}{5} g(k) \mu^i \leq g(k) \mu^i,$$

as desired. □

Remark Recall the initial question from Figure 10.4, which asked whether the clustering effect wins over the entropy effect or not. This question enabled us to motivate the setup of the proof but in the end, we did not specifically address it. Notice that the above proof in fact solves the problem (see Exercise 10.3).

10.4 Back to the spectrum: An exposition of the proof

10.4.1 Heuristic explanation

Let us now apply the strategy we developed for $\mathcal{T}_n$ to the case of the spectral sample $\mathcal{S}_{f_n}$. Our goal is to prove Theorem 10.3 (of which Theorems 10.1 and 10.2 can be shown to follow). Let $\mathcal{S}_{f_n} \subset [0,n]^2$ be our spectral sample. We have seen (Theorem 9.8) that with positive probability $|\mathcal{S}_{f_n}| \asymp n^2 \alpha_4(n)$. For all $1 < u < n^2 \alpha_4(n)$, we wish to understand the probability $\hat{\mathbb{P}}[0 < |\mathcal{S}_{f_n}| < u]$. Following the notations we used for $\mathcal{T}_n$, let $\mathcal{S}_{f_n}^{|u}$ be the spectral sample conditioned on the event $\{0 < |\mathcal{S}_{f_n}| < u\}$.

Question: How does $\mathcal{S}_{f_n}^{|u}$ typically look?

To answer this question, one has to understand whether $\mathcal{S}_{f_n}^{|u}$ tends to be *localized* or not. Recall from Figure 10.4 the illustration of the competition between entropy and clustering effects in the case of $\mathcal{T}_n$. The same figure applies to the spectral sample $\mathcal{S}_{f_n}$. We later state a **clustering lemma** (Lemma 10.18) that will strongly support the localized behavior described in the next proposition.

Therefore we are guessing that our conditional set $\mathcal{S}_{f_n}^{|u}$ will tend to localize into $O(1)$ many squares of a certain scale r and will have a "normal" size within these r-squares. It remains to understand what this mesoscopic scale r as a function of u is.

By "scale invariance", one expects that if $\mathcal{S}_{f_n}$ is conditioned to live in a square of size r, then $|\mathcal{S}_{f_n}|$ will be of order $r^2 \alpha_4(r)$ with positive conditional probability. More precisely, the following lemma will be proved in Problem 10.6.

Lemma 10.12   *There is a universal $c \in (0,1)$ such that for any n and for any r-square $B \subset [n/4, 3n/4]^2$ in the "bulk" of $[0,n]^2$, one has*

$$\hat{\mathbb{P}}\Big[\frac{|\mathcal{S}_{f_n}|}{r^2 \alpha_4(r)} \in (c, 1/c) \,\Big|\, \mathcal{S}_{f_n} \neq \emptyset \text{ and } \mathcal{S}_{f_n} \subset B\Big] > c. \qquad (10.8)$$

In fact this lemma holds uniformly in the position of the r-square B inside $[0,n]^2$, but we do not discuss this here.

What this lemma tells us is that for any $1 < u < n^2 \alpha_4(n)$, if one chooses $r = r_u$ in such a way that $r^2 \alpha_4(r) \asymp u$, then we expect to have the following estimate:

$$\hat{\mathbb{P}}[0 < |\mathcal{S}_{f_n}| < u] \asymp \hat{\mathbb{P}}[\mathcal{S}_{f_n} \text{ intersects } O(1) \text{ } r\text{-squares in } [0,n]^2]$$
$$\asymp \hat{\mathbb{P}}[\mathcal{S}_{f_n} \text{ intersects a single } r\text{-square in } [0,n]^2]$$

At this point, let us introduce a concept that will be very helpful in what follows.

Definition 10.13 ("*r*-smoothing")   Let $1 \le r \le n$. Consider the domain $[0,n]^2$ and divide it into a grid of squares of edge-length r. (If $1 \ll r \ll n$, one can view this grid as a *mesoscopic* grid).

If n is not divisible by r, write $n = mr + q$ and consider the grid of r-squares covering $[0, (m+1)r]^2$.

Now, for each subset $S \subset [0,n]^2$, define $S_{(r)}$ to be the set of $r \times r$ squares in the above grid that intersect S. In particular $|S_{(r)}|$ will correspond to the number of such r-squares which intersect S. With a slight abuse of notation, $S_{(r)}$ will sometimes also denote the actual subset of $[0,n]^2$ consisting of the union of these r-squares.

One can view the application $S \mapsto S_{(r)}$ as an *r*-**smoothing** because all of the details below the scale r are lost.

Remark   Note that in Definition 10.8, we relied on a slightly different notion of "*r*-smoothing" because in that case, $\mathcal{T}_{(r)}$ could also include r-branches that might go extinct by the time they reached scale one. The advantage of this choice was that there was an exact scale-invariance from $\mathcal{T}$ to $\mathcal{T}_{(r)}$ while in the case of $\mathcal{S}_{f_n}$, there is no such exact scale-invariance from $\mathcal{S}$ to $\mathcal{S}_{(r)}$.

With these notations, the preceding discussion leads us to believe that the following proposition should hold.

Proposition 10.14   *For all $1 \le r \le n$, one has*

$$\hat{\mathbb{P}}[0 < |\mathcal{S}_{f_n}| < r^2 \alpha_4(r)] \asymp \hat{\mathbb{P}}_{f_n}[|\mathcal{S}_{(r)}| = 1].$$

Before explaining the setup used in (GPS10) to prove such a result, let us check that it indeed implies Theorem 10.3. By neglecting the boundary issues, one has

$$\hat{\mathbb{P}}_{f_n}[|\mathcal{S}_{(r)}| = 1] \asymp \sum_{\substack{r\text{-squares} \\ B \subset [n/4, 3n/4]^2}} \hat{\mathbb{P}}[\mathcal{S}_{f_n} \ne \emptyset \text{ and } \mathcal{S}_{f_n} \subset B]. \tag{10.9}$$

There are $O(\frac{n^2}{r^2})$ such B squares, and for each of these, one can check (see Exercise 10.5) that

$$\hat{\mathbb{P}}[\mathcal{S}_{f_n} \ne \emptyset \text{ and } \mathcal{S}_{f_n} \subset B] \asymp \alpha_4(r,n)^2. \tag{10.10}$$

Therefore, Proposition 10.14 indeed implies Theorem 10.3.

10.4.2 Setup and organization of the proof of Proposition 10.14

To start with, assume we knew that disjoint regions in the spectral sample $\mathscr{S}_{f_n}$ behave more or less independently of each other in the following (vague) sense. For any $k \geq 1$ and any mesoscopic scale $1 \leq r \leq n$, if one conditions on $\mathscr{S}_{(r)}$ to be equal to $B_1 \cup \cdots \cup B_k$ for k disjoint r-squares, then the conditional law of $\mathscr{S}_{|\cup B_i}$ should be "similar" to an independent product of $\mathcal{L}[\mathscr{S}_{|B_i} \mid \mathscr{S} \cap B_i \neq \emptyset]$, $i \in \{1, \ldots, k\}$. Similarly as in the tree case (where the analogous property for $\mathcal{T}_n$ was an exact independence factorization), and assuming that the above comparison with an independent product could be made quantitative, this would potentially imply the following upper bound for a certain absolute constant $c > 0$:

$$\hat{\mathbb{P}}[0 < |\mathscr{S}_{f_n}| < r^2 \alpha_4(r)] \leq \sum_{k \geq 1} \hat{\mathbb{P}}[|\mathscr{S}_{(r)}| = k](1 - c)^k. \qquad (10.11)$$

This means that even if one managed to obtain a good control on the dependency structure within $\mathscr{S}_{f_n}$ (in the above sense), one would still need to have a good estimate on $\hat{\mathbb{P}}[|\mathscr{S}_{(r)}| = k]$ to deduce Proposition 10.14. This part of the program is achieved in (GPS10) without requiring any information on the dependency structure of $\mathscr{S}_{f_n}$. More precisely, the following result is proved:

Theorem 10.15 (GPS10) *There is a subexponential function $g \mapsto g(k)$, such that for any $1 \leq r \leq n$ and any $k \geq 1$,*

$$\hat{\mathbb{P}}[|\mathscr{S}_{(r)}| = k] \leq g(k)\, \hat{\mathbb{P}}[|\mathscr{S}_{(r)}| = 1].$$

The proof of this result is described briefly in the next subsection.

One can now describe how the proof of Theorem 10.3 is organized in (GPS10). It is divided into three main parts:

1. The first part deals with proving the multiscale subexponential bound on the lower-tail of $|\mathscr{S}_{(r)}|$ given by Theorem 10.15.
2. The second part consists in proving as much as we can on the dependency structure of $\mathscr{S}_{f_n}$. Unfortunately here, it seems to be very challenging to achieve a good understanding of all the "independence" that should be present within $\mathscr{S}_{f_n}$. The only hint of independence which was finally proved in (GPS10) is a very *weak* one (see Subsection 10.4.4). In particular, it is too weak to readily imply a bound like (10.11).
3. As disjoint regions of the spectral sample $\mathscr{S}_{f_n}$ are not known to behave independently of each other, the third part of the proof consists in adapting the setup we used for the tree (where distinct branches evolve

exactly independently of each other) into a setup in which the weak hint of independence obtained in the second part of the program turns out to be enough to imply the bound given by (10.11) for an appropriate absolute constant $c > 0$. This final part of the proof will be discussed in Subsection 10.4.5.

The next three subsections will be devoted to each of these three parts of the program.

10.4.3 Some words about the subexponential bound on the lower tail of $\mathscr{S}_{(r)}$

In this subsection, we turn our attention to the proof of the first part of the program, that is, on Theorem 10.15. In fact, as in the case of $\mathcal{T}_n$, the following more explicit statement is proved in (GPS10).

Theorem 10.16 (GPS10) *There exists an absolute constant $\theta > 0$ such that for any $1 \le r \le n$ and any $k \ge 1$,*

$$\hat{\mathbb{P}}\big[|\mathscr{S}_{(r)}| = k\big] \le 2^{\theta \log_2^2(k+2)}\, \hat{\mathbb{P}}\big[|\mathscr{S}_{(r)}| = 1\big].$$

Remark Note that the theorems from (BKS99) on the noise sensitivity of percolation are all particular cases ($r = 1$) of this intermediate result in (GPS10).

The main idea in the proof of this theorem is in some sense to assign a *tree structure* to each possible set $\mathscr{S}_{(r)}$. The advantage of working with a tree structure is that it is easier to work with inductive arguments. In fact, once a mapping $\mathscr{S}_{(r)} \mapsto$ "tree structure" has been designed, the proof proceeds similarly as in the case of $\mathcal{T}_{(r)}$ by double induction on the depth of the tree as well as on $k \ge 1$. Of course, this mapping is a delicate affair: it has to be designed in an "efficient" way so that it can compete against entropy effects caused by the exponential growth of the number of tree structures.

We do not give the details of how to define such a mapping, but let us describe informally how it works. More specifically than a tree structure, we will in fact assign an *annulus structure* to each set $\mathscr{S}_{(r)}$.

Definition 10.17 Let $\mathcal{A}$ be a finite collection of disjoint (topological) annuli in the plane. We call this an **annulus structure**. Furthermore, we will say that a set $S \subset \mathbb{R}^2$ is **compatible** with $\mathcal{A}$ (or vice versa) if it is contained in $\mathbb{R}^2 \setminus \bigcup \mathcal{A}$ and intersects the inner disk of each annulus in $\mathcal{A}$. Note that it is allowed that one annulus is "inside" of another annulus.

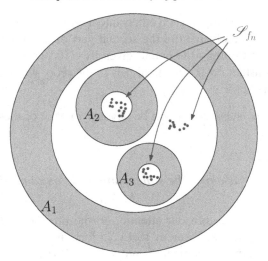

Figure 10.5 An example of an **annulus structure**
$\mathcal{A} := \{A_1, A_2, A_3\}$ compatible with a spectral sample $\mathscr{S}_{f_n}$.

The mapping procedure in (GPS10) assigns to each $\mathscr{S}_{(r)}$ an annulus structure $\mathcal{A} \subset [0,n]^2$ in such a way that it is compatible with $\mathscr{S}_{(r)}$. See Figure 10.5 for an example. Again, we do not describe this procedure nor discuss the obvious boundary issues which arise here, but let us state a crucial property satisfied by annulus structures.

Lemma 10.18 (clustering Lemma) *If $\mathcal{A}$ is an annulus structure contained in $[0,n]^2$, then*

$$\hat{\mathbb{P}}[\mathscr{S}_{(r)} \text{ is compatible with } \mathcal{A}] \leq \prod_{A \in \mathcal{A}} \alpha_4(A)^2 \,,$$

where $\alpha_4(A)$ denotes the probability of having a four-arm event in the annulus A.

Remark To deal with boundary issues, one would also need to incorporate within our annulus structures half-annuli centered on the boundary as well as quarter disks centered at the corners of $[0,n]^2$.

Let us briefly comment on this lemma.

- First of all, its proof is an elegant combination of linear algebra and percolation. It is a short and relatively elementary argument. See Lemma 4.3 in (GPS10).

- It is very powerful in dealing with the possible non-injectivity of the mapping $\mathcal{S}_{(r)} \mapsto \mathcal{A}$. Indeed, while describing the setup above, one might have objected that if the mapping were not injective enough, then the cardinality of the "fibers" above each annulus structure would have to be taken into account as well. Fortunately, the above lemma reads as follows: for any fixed annulus structure $\mathcal{A}$,

$$\sum_{\mathcal{S}_{(r)}:\mathcal{S}_{(r)} \mapsto \mathcal{A}} \hat{\mathbb{P}}[\mathcal{S}_{(r)}] \le \hat{\mathbb{P}}[\mathcal{S}_{(r)} \text{ is compatible with } \mathcal{A}] \le \prod_{A \in \mathcal{A}} \alpha_4(A)^2.$$

- Another essential feature of this lemma is that it quantifies very efficiently the fact that the clustering effect wins over the entropy effect in the sense of Figure 10.4. The mechanism responsible for this is that the probability of the *four-arm* event squared has an exponent (equal to $5/2$ on $\mathbb{T}$) larger than the *volume* exponent equal to 2. To illustrate this, let us analyze the situation when $k = 2$ (still neglecting boundary issues). The probability that the spectrum $\mathcal{S}_{f_n}$ intersects two and only two r-squares at macroscopic distance $\Omega(n)$ from each other can be easily estimated using the lemma. Indeed, in such a case, $\mathcal{S}_{(r)}$ would be compatible with an annulus structure consisting of two annuli, each being approximately of the type $A(r, n)$. There are $O(\frac{n^2}{r^2}) \times O(\frac{n^2}{r^2})$ such possible annulus structures. Using the lemma each of them costs (on $\mathbb{T}$) $(\frac{r}{n})^{5+o(1)}$. An easy exercise shows that this is much smaller than $\hat{\mathbb{P}}[|\mathcal{S}_{(r)}| = 2]$. In other words, if $|\mathcal{S}_{(r)}|$ is conditioned to be small, it tends to be localized. Also, the way that the lemma is stated makes it very convenient to work with higher values of k.

The details of the proof of Theorem 10.16 can be found in (GPS10). The double induction there is in some sense very close to the one we carried out in detail in Subsection 10.3.4 in the case of the tree; this is the reason why we included this latter proof. For those who might read the proof in (GPS10), there is a notion of *overcrowded cluster* defined there; it exactly corresponds in the case of the tree to stopping the analysis above scale J_k instead of going all the way to scale 1 (note that without stopping at this scale J_k, the double induction in subsection 10.3.4 would have failed).

10.4.4 Some words on the weak independence property proved in (GPS10)

This part of the program is in some sense the main one. To introduce it, let us start by a naive but tempting strategy. What the first part of the program

(Theorem 10.16) tells us is that for any mesoscopic scale $1 \leq r \leq n$, if $\mathscr{S}_{f_n}$ is nonempty, it is very unlikely that it will intersect few squares of size r. In other words, it is very unlikely that $|\mathscr{S}_{(r)}|$ will be small. Let $B_1, \ldots, B_m$ denote the set of $O(n^2/r^2)$ r-squares that tile $[0,n]^2$. One might try the following *scanning procedure*: explore the spectral sample $\mathscr{S}_{f_n}$ inside the squares B_i one at a time. More precisely, before starting the scanning procedure, we consider our spectral sample $\mathscr{S}_{f_n}$ as a random subset of $[0,n]^2$ about which we do not know anything yet. Then, at step 1, we reveal $\mathscr{S}_{|B_1}$. This gives us some partial information about $\mathscr{S}_{f_n}$. What we still have to explore is a random set of $[0,n]^2 \setminus B_1$ that follows the law of a spectral sample conditioned on what was seen in B_1 and we keep going in this way. By Theorem 10.16, many of these squares will be nonempty. Now, it is not hard to prove the following lemma (using similar methods as in Problem 10.6).

Lemma 10.19 *There is a universal constant $c > 0$ such that for any r-square B in the bulk $[n/4, 3n/4]^2$, one has*

$$\hat{\mathbb{P}}[|\mathscr{S}_{f_n} \cap B| > c\, r^2 \alpha_4(r) \mid \mathscr{S}_{f_n} \cap B \neq \emptyset] > c.$$

This lemma in fact holds uniformly in the position of B inside $[0,n]^2$.

If one could prove the following (much) stronger result: there exists a universal constant $c > 0$ such that uniformly on the sets $S \subset [0,n]^2 \setminus B$ one has

$$\hat{\mathbb{P}}[|\mathscr{S}_{f_n} \cap B| > c\, r^2 \alpha_4(r) \mid \mathscr{S}_{f_n} \cap B \neq \emptyset \text{ and } \mathscr{S}_{|B^c} = S] > c, \tag{10.12}$$

then it would not be hard to make the above scanning strategy work together with Theorem 10.16 to obtain Theorem 10.3. (Note that such a result would indeed give a strong hint of independence within $\mathscr{S}_{f_n}$.) However, as we discussed before, the current understanding of the independence within $\mathscr{S}_{f_n}$ is far from giving such a statement. Instead, the following result is proved in (GPS10). We provide here a slightly simplified version.

Theorem 10.20 (GPS10) *There exists a uniform constant $c > 0$ such that for any set $W \subset [0,n]^2$ and any r-square B such that $B \cap W = \emptyset$, one has*

$$\hat{\mathbb{P}}[|\mathscr{S}_{f_n} \cap B| > c\, r^2 \alpha_4(r) \mid \mathscr{S}_{f_n} \cap B \neq \emptyset \text{ and } \mathscr{S}_{f_n} \cap W = \emptyset] > c.$$

Note that this theorem in some sense interpolates between part of Lemma 10.12 and Lemma 10.19 which correspond respectively to the special cases $W = B^c$ and $W = \emptyset$. Yet it looks very weak compared to the expected (10.12) which is stated uniformly on the behavior of $\mathscr{S}_{f_n}$ outside of B.

A crucial step in the proof of Theorem 10.20 is to understand the following "one-point function" for any $x \in B$ at distance at least $r/3$ from the boundary:

$$\hat{\mathbb{P}}[x \in \mathscr{S}_{f_n} \text{ and } \mathscr{S}_{f_n} \cap W = \emptyset].$$

Note that by Proposition 9.11, this probability has a very nice description in terms of an explicit coupling of two i.i.d. percolation configurations and the notion of pivotality. If (ω_1, ω_2) is a coupling of two i.i.d. percolation realizations on $[0, n]^2$ which are such that

$$\begin{cases} \omega_1 = \omega_2 & \text{on } W^c \\ \omega_1, \omega_2 & \text{are independent on } W \end{cases}$$

then, when $x \notin W$, we have

$$\hat{\mathbb{P}}[x \in \mathscr{S}_{f_n} \text{ and } \mathscr{S}_{f_n} \cap W = \emptyset] = \mathbb{P}[x \text{ is pivotal for } \omega_1 \text{ and } \omega_2].$$

Thanks to this, the proof of Theorem 10.20 proceeds by analyzing this W-coupling. See (GPS10) for the complete details.

10.4.5 Adapting the setup to the weak hint of independence

Assuming the weak hint of independence given by Theorem 10.20, it seems we are in bad shape if we try to apply the above naive sequential scanning procedure. Indeed, we face the following two obstacles:

1. The first obstacle is that one would keep a good control only as far as one would not see any "spectrum." Namely, while revealing $\mathscr{S}_{|B_i}$ one at a time, the first time one finds a square B_i such that $\mathscr{S}_{|B_i} \neq \emptyset$, one would be forced to stop the scanning procedure there. In particular, if the size of the spectrum in this first nontrivial square does not exceed $r^2 \alpha_4(r)$, then we cannot conclude anything.
2. The second obstacle is that, besides the conditioning $\mathscr{S} \cap W = \emptyset$, our estimate is also conditioned on the event that $\mathscr{S} \cap B \neq \emptyset$. In particular, in the above "naive" scanning strategy where squares are revealed in a sequential way, at each step one would have to update the probability that $\mathscr{S} \cap B_{i+1} \neq \emptyset$ based on what was discovered so far.

It is the purpose of this third part of the program to adapt the above scanning strategy to these constraints.

Let us start with the first obstacle. Assume that we scan the domain $[0, n]^2$ in a sequential way, that is, we choose an increasing family of subsets $(W_l)_{l \geq 1} = (\{w_1, \ldots, w_l\})_{l \geq 1}$. At each step, we reveal what $\mathscr{S}_{|\{w_{l+1}\}}$ is, conditioned on what was discovered so far (i.e., conditioned on $\mathscr{S}_{|W_l}$). From the weak independence Theorem 10.20, it is clear that if we want this strategy to have any chance to be successful, we have to choose $(W_l)_{l \geq 1}$ in such a way that $(\mathscr{S}_{f_n} \cap W_l)_{l \geq 1}$ will remain empty for some time (so that we can continue to rely on our weak independence result); of course this cannot remain empty forever, so the game is to choose the increasing family $(W_l)_{l \geq 1}$ in such a way that the first time $\mathscr{S}_{f_n} \cap \{w_l\}$ will happen to be nonempty, it should give a strong indication that $\mathscr{S}_{f_n}$ is large in the r-neighborhood of w_l.

As we have seen, revealing the entire mesoscopic boxes B_i one at a time is not a successful idea. Here is a much better idea (which is not yet the right one due to the second obstacle, but we are getting close): in each r-square B_i, instead of revealing all the bits, let us reveal only a very small proportion δ_r of them. Lemma 10.19 tells us that if $\mathscr{S} \cap B_i \neq \emptyset$, then each point $x \in B_i$ has probability of order $\alpha_4(r)$ to be in $\mathscr{S}_{f_n}$. Therefore if we choose $\delta_r \ll (r^2 \alpha_4(r))^{-1}$, then with high probability, by revealing only a proportion δ_r of the points in B_i, we will "miss" the spectral sample $\mathscr{S}_{f_n}$. Hence, we have to choose $\delta_r \geq (r^2 \alpha_4(r))^{-1}$. In fact choosing $\delta \asymp (r^2 \alpha_4(r))^{-1}$ is exactly the right balance. Indeed, we know from Theorem 10.16 that many r-squares B_i will be touched by the spectral sample; now, in this more sophisticated scanning procedure, if the first such square encountered happens to contain few points (i.e. $\ll r^2 \alpha_4(r)$), then with the previous scanning strategy, we would "lose," but with the present one, owing to our choice of δ_r, most likely we will keep $\mathscr{S}_{f_n} \cap W_l = \emptyset$ so that we can continue further on until we reach a "good" square (i.e. a square containing of order $r^2 \alpha_4(r)$ points).

Now, Theorems 10.16 and 10.20 together tell us that with high probability, one will eventually reach such a good square. Indeed, suppose the m first r-squares touched by the spectral sample happened to contain few points; then, most likely, if W_{l_m} is the set of bits revealed so far, by our choice of δ_r we will still have $\mathscr{S} \cap W_{l_m} = \emptyset$. This allows us to still rely on Theorem 10.20, which basically tells us that there is a positive conditional probability for the next one to be a "good" square (we are neglecting the second obstacle here). This says that the probability to visit m consecutive bad squares seems to decrease exponentially fast. Because m is typically very large (by Theorem 10.16), we conclude that, with high probability, we will finally reach good squares. In the first good square encountered, by our choice of δ_r, there is now a positive probability to reveal a bit present in $\mathscr{S}_{f_n}$. In this case, the sequential scanning will have to stop, as we will not be able to use our weak independence result anymore, but this is not a big issue: indeed, assume that you have some random set $S \subset B$. If by revealing each bit only with probability δ_r, you end up finding a point in S, most likely your set S is at least of size $\Omega(r^2\alpha_4(r))$. This is exactly the size we are looking for in Theorem 10.3.

Now, only the second obstacle remains. It can be rephrased as follows: assume you applied the above strategy in $B_1,\ldots,B_h$ (i.e., you revealed each point in B_i, $i \in \{1,\ldots,h\}$ only with probability δ_r) and that you did not find any spectrum yet. In other words, if W_l denotes the set of points visited so far, then $\mathscr{S}_{f_n} \cap W_l = \emptyset$. Now if B_{h+1} is the next r-square to be scanned (still in a "dilute" way with intensity δ_r), we seem to be in good shape because we know how to control the conditioning $\mathscr{S}_{f_n} \cap W_l = \emptyset$. However, if we want to rely on the uniform control given by Theorem 10.20, we also need to further condition on $\mathscr{S}_{f_n} \cap B_{h+1} \neq \emptyset$. In other words, we need to control the following conditional expectation:

$$\hat{\mathbb{P}}[\mathscr{S}_{f_n} \cap B_{h+1} \neq \emptyset \mid \mathscr{S}_{f_n} \cap W_l = \emptyset].$$

It is quite involved to estimate such quantities. Fortunately, by changing our sequential scanning procedure into a slightly more "abstract" procedure, one can avoid dealing with such terms. More precisely, within each r-square B, we will still reveal only a δ_r proportion of the bits (so that the first obstacle is still taken care of), but instead of operating in a sequential way (i.e., scanning B_1, then B_2 and so on), we will gain a lot by considering the combination of Theorem 10.16 and Theorem 10.20 in a more abstract fashion. Namely, the following large deviation lemma from (GPS10) captures exactly what we need in our present situation.

Lemma 10.21 (GPS10)   *Let $X_i, Y_i \in \{0,1\}, i \in \{1,\ldots,m\}$ be random variables such that for each i $Y_i \leq X_i$ a.s. If $\forall J \subset [m]$ and $\forall i \in [m] \setminus J$, we have*

$$\mathbb{P}[Y_i = 1 \mid Y_j = 0, \forall j \in J] \geq c\,\mathbb{P}[X_i = 1 \mid Y_j = 0, \forall j \in J], \quad (10.13)$$

then if $X := \sum X_i$ and $Y := \sum Y_i$, one has that

$$\mathbb{P}[Y = 0 \mid X > 0] \leq c^{-1}\mathbb{E}[e^{-(c/e)X} \mid X > 0].$$

Recall that $B_1,\ldots,B_m$ denotes the set of r-squares that tile $[0,n]^2$. For each $i \in [m]$, let $X_i := 1_{\mathscr{S} \cap B_i \neq \emptyset}$ and $Y_i := 1_{\mathscr{S} \cap B_i \cap W \neq \emptyset}$, where $\mathcal{W}$ is an independent uniform random subset of $[0,n]^2$ of intensity δ_r.

This lemma enables us to combine our two main results, Theorems 10.20 and 10.16, in a very nice way: By our choice of the intensity δ_r, Theorem 10.20 exactly states that the assumption (10.13) is satisfied for a certain constant $c > 0$. Lemma 10.21 then implies that

$$\hat{\mathbb{P}}[Y = 0 \mid X > 0] \leq c^{-1}\mathbb{E}[e^{-(c/e)X} \mid X > 0].$$

Now, notice that $X = \sum X_i$ exactly corresponds to $|\mathscr{S}_{(r)}|$ while the event $\{X > 0\}$ corresponds to $\{\mathscr{S}_{f_n} \neq \emptyset\}$ and the event $\{Y = 0\}$ corresponds to $\{\mathscr{S}_{f_n} \cap \mathcal{W} = \emptyset\}$. Therefore Theorem 10.16 leads us to

$$\hat{\mathbb{P}}[\mathscr{S}_{f_n} \cap \mathcal{W} = \emptyset, \mathscr{S}_{f_n} \neq \emptyset] \leq c^{-1}\mathbb{E}[e^{-(c/e)|\mathscr{S}_{(r)}|}, \mathscr{S}_{f_n} \neq \emptyset]$$

$$\leq c^{-1}\sum_{k\geq 1}\hat{\mathbb{P}}[|\mathscr{S}_{(r)}| = k]e^{-(c/e)k}$$

$$\leq c^{-1}\Big(\sum_{k\geq 1}2^{\theta\log_2^2(k+2)}e^{-(c/e)k}\Big)\hat{\mathbb{P}}[|\mathscr{S}_{(r)}| = 1]$$

$$\leq C(\theta)\hat{\mathbb{P}}[|\mathscr{S}_{(r)}| = 1] \asymp \frac{n^2}{r^2}\alpha_4(r,n)^2, \qquad (10.14)$$

where (10.10) is used in the last step.

This shows that on the event that $\mathscr{S}_{f_n} \neq \emptyset$, it is very unlikely that we do not detect the spectral sample on the δ_r-dilute set $\mathcal{W}$. This is enough for us to conclude using the following identity:

$$\hat{\mathbb{P}}[\mathscr{S}_{f_n} \cap \mathcal{W} = \emptyset \mid \mathscr{S}_{f_n}] = (1 - \delta_r)^{|\mathscr{S}_{f_n}|} = (1 - \frac{1}{r^2\alpha_4(r)})^{|\mathscr{S}_{f_n}|}.$$

Indeed, by averaging this identity we obtain

$$\hat{\mathbb{P}}[\mathscr{S}_{f_n} \cap \mathcal{W} = \emptyset, \mathscr{S}_{f_n} \neq \emptyset] = \hat{\mathbb{E}}[\hat{\mathbb{P}}[\mathscr{S}_{f_n} \cap \mathcal{W} = \emptyset \mid \mathscr{S}_{f_n}]1_{\mathscr{S}_{f_n}\neq\emptyset}]$$

$$= \hat{\mathbb{E}}[(1 - \frac{1}{r^2\alpha_4(r)})^{|\mathscr{S}_{f_n}|}1_{\mathscr{S}_{f_n}\neq\emptyset}]$$

$$\geq \Omega(1)\hat{\mathbb{P}}[0 < |\mathscr{S}_{f_n}| < r^2\alpha_4(r)],$$

which, combined with (10.14), yields the desired upper bound in Theorem 10.3. See Problem 10.7 for the lower bound.

10.5 The radial case

The next chapter focuses on the existence of *exceptional times* in the model of dynamical percolation. A main tool in the study of these exceptional times is the spectral measure $\hat{\mathbb{Q}}_{g_R}$ where g_R is the Boolean function $g_R : \{-1,1\}^{O(R^2)} \to \{0,1\}$ defined to be the indicator function of the one-arm event $\{0 \longleftrightarrow \partial B(0,R)\}$. Note that by definition, g_R is such that $\|g_R\|_2^2 = \alpha_1(R)$.

In (GPS10), the following "sharp" theorem on the lower tail of $\mathscr{S}_{g_R}$ is proved.

Theorem 10.22 (GPS10) *Let g_R be the one-arm event in $B(0,R)$. Then for any $1 \leq r \leq R$, one has*

$$\hat{\mathbb{Q}}_{g_R}[0 < |\mathscr{S}_{g_R}| < r^2 \alpha_4(r)] \asymp \frac{\alpha_1(R)^2}{\alpha_1(r)}. \tag{10.15}$$

The proof of this theorem is in many ways similar to the chordal case (Theorem 10.3). An essential difference is that the "clustering vs. entropy" mechanism is very different in this case. Indeed in the chordal left to right case, when $\mathscr{S}_{f_n}$ is conditioned to be very small, the proof of Theorem 10.3 shows that typically $\mathscr{S}_{f_n}$ localizes in some r-square whose location is "uniform" in the domain $[0,n]^2$. In the radial case, the situation is very different: $\mathscr{S}_{g_R}$ conditioned to be very small will in fact tend to localize in the r-square centered at the origin. This means that the analysis of the mesoscopic behavior (i.e., the analog of Theorem 10.16) has to be adapted to the radial case. In particular, in the definition of an annulus structure, the annuli containing the origin play a distinguished role. See (GPS10) for complete details.

10.6 Exercises

10.1 Prove Proposition 10.5.

10.2 Consider the fractal percolation process $\mathcal{T}^i$, $i \geq 1$ introduced in this chapter. (Recall that $\mathcal{T}_{2^i} \equiv \mathcal{T}^i$). Recall that in Section 10.3, it was important to estimate the quantity $\mathbb{P}[|\mathcal{T}^i| = 1]$. This is one of the purposes of the present exercise.

(a) Let $p_i := \mathbb{P}[|\mathcal{T}^i| = 1]$. By recursion, show that there is a constant $c \in (0, 1)$ so that, as $i \to \infty$

$$p_i \sim c\mu^i,$$

where $\mu := 4p(1 - p + pq)^3$ and q is the probability of extinction for the Galton–Watson tree corresponding to $(\mathcal{T}^i)_{i \geq 1}$.

(b) Using the generating function $s \mapsto f(s)(= E(s^{\text{ number of offspring}})$ of this Galton–Watson tree, and by studying the behavior of its ith iterates $f^{(i)}$, prove the same result with $\mu := f'(q)$. Check that it gives the same formula.

(c) Recall the definition of $p_{m,b}$ from Section 10.3. Let $p_{m,\infty}$ be the probability that exactly one person at generation m survives forever. Prove that

$$p_{m,\infty} = (1 - q)\mu^m$$

for the same exponent μ. Prove Lemma 10.11. Finally, prove that

$$\lim_{b \to \infty} p_{m,b} = p_{m,\infty}.$$

10.3 Extract from the proof of Lemma 10.10 the answer to the question asked in Figure 10.4.

10.4 Prove that

$$\text{Theorem } 10.3 \Rightarrow \text{Theorem } 10.2 \Rightarrow \text{Theorem } 10.1$$

10.5 Consider an r-square $B \subset [n/4, 3n/4]^2$ in the "bulk" of $[0, n]^2$.

(a) Prove using Proposition 9.5 that

$$\hat{\mathbb{P}}[\mathscr{S}_{f_n} \neq \emptyset \text{ and } \mathscr{S}_{f_n} \subset B] \asymp \alpha_4(r, n)^2$$

(b) Check that the clustering Lemma 10.18 is consistent with this estimate.

10.6 The purpose of this exercise is to prove Lemma 10.12.

(a) Using Proposition 9.5, prove that for any $x \in B$ at distance $r/3$ from the boundary of B,

$$\mathbb{P}[x \in \mathscr{S}_{f_n} \text{ and } \mathscr{S}_{f_n} \cap B^c = \emptyset] \asymp \alpha_4(r)\alpha_4(r, n)^2.$$

(b) Recover the same result using Proposition 9.6 instead.
(c) Conclude using Exercise 10.5 that $\hat{\mathbb{E}}[|\mathscr{S}_{f_n} \cap \bar{B}| \,|\, \mathscr{S}_{f_n} \neq \emptyset \text{ and } \mathscr{S}_{f_n} \subset B] \asymp r^2\alpha_4(r)$, where $\bar{B} \subset B$ is the set of points $x \in B$ at distance at least $r/3$ from the boundary.
(d) Study the second-moment $\hat{\mathbb{E}}[|\mathscr{S}_{f_n} \cap \bar{B}|^2 \,|\, \mathscr{S}_{f_n} \neq \emptyset \text{ and } \mathscr{S}_{f_n} \subset B]$.
(e) Deduce Lemma 10.12.

10.7 Most of this chapter was devoted to the explanation of the proof of Theorem 10.3. Note that we in fact only discussed how to prove the upper bound. This is because the lower bound is much easier to prove and this is the purpose of this problem.

(a) Deduce from Lemma 10.12 and Exercise 10.5(a) that the lower bound on
$\hat{\mathbb{P}}[0 < |\mathscr{S}_{f_n}| < r^2\alpha_4(r)]$ given in Theorem 10.3 is correct; that is, show that there exists a constant $c > 0$ such that

$$\hat{\mathbb{P}}[0 < |\mathscr{S}_{f_n}| < r^2\alpha_4(r)] > c\frac{n^2}{r^2}\alpha_4(r,n)^2.$$

(b) (Hard) In the same fashion, prove the lower bound part of Theorem 10.22.

11

Applications to dynamical percolation

In this section, we present a very natural model where percolation undergoes a time-evolution: this is the model of **dynamical percolation** described below. The study of the "dynamical" behavior of percolation as opposed to its "static" behavior turns out to be very rich: interesting phenomena arise especially at the phase transition point. We will see that in some sense, dynamical planar percolation at criticality is a very unstable or chaotic process. In order to understand this instability, sensitivity of percolation (and therefore its Fourier analysis) will play a key role. In fact, the original motivation for the paper (BKS99) on noise sensitivity was to solve a particular problem in the subject of dynamical percolation. (Ste09) provides a recent survey on the subject of dynamical percolation.

We mention that one can read all but the last section of the present chapter without having read Chapter 10.

11.1 The model of dynamical percolation

This model was introduced by Häggström, Peres, and Steif (HPS97) inspired by a question that Paul Malliavin asked at a lecture at the Mittag-Leffler Institute in 1995. This model was invented independently by Itai Benjamini.

In the general version of this model as it was introduced, given an arbitrary graph G and a parameter p, the edges of G switch back and forth according to independent two-state continuous time Markov chains where closed switches to open at rate p and open switches to closed at rate $1 - p$. Clearly, the product measure with density p, denoted by π_p in this chapter, is the unique stationary distribution for this Markov process. The general question studied in dynamical percolation is whether, when we start with the stationary distribution π_p, there exist atypical times at which the percolation structure looks markedly different than that at a fixed time. In almost all cases, the term "markedly different" refers to the existence or nonexistence of an infinite connected component. Dynamical percolation on site percolation models, which includes our most important case of the hexagonal lattice, is defined analogously.

We very briefly summarize a few early results in the area. It was shown in (HPS97) that below criticality, there are no times at which there is an infinite cluster and above criticality, there is an infinite cluster at all times. See the exercises. In (HPS97), examples of graphs that do not percolate at criticality but for which there exist exceptional times where percolation occurs were given. (Also given were examples of graphs that do percolate at criticality but for which there exist exceptional times where percolation does not occur.) A fairly refined analysis of the case of so-called *spherically symmetric* trees was given. See the exercises for some of these.

Given the above results, it is natural to ask what happens on the standard graphs that we work with. Recall that for $\mathbb{Z}^2$, we have seen that there is no percolation at criticality. It turns out that it is also known (see later) that for $d \geq 19$, there is no percolation at criticality for $\mathbb{Z}^d$. It is a major open question to prove that this is also the case for intermediate dimensions; the consensus is that this should be the case.

11.2 What's going on in high dimensions: $\mathbb{Z}^d, d \geq 19$?

For the high-dimensional case, $\mathbb{Z}^d, d \geq 19$, it was shown in (HPS97) that there are no exceptional times of percolation at criticality.

Theorem 11.1 (HPS97) *For the integer lattice $\mathbb{Z}^d$ with $d \geq 19$, dynamical critical percolation has no exceptional times of percolation.*

The key reason for this is a highly nontrivial result due to work of Hara and Slade (HS94), using earlier work of Barsky and Aizenman (BA91), which says that if $\theta(p)$ is the probability that the origin percolates when the parameter is p, then for $p \geq p_c$

$$\theta(p) = O(p - p_c). \tag{11.1}$$

(This implies in particular that there is no percolation at criticality.) In fact, this is the only thing that is used in the proof and hence the result holds whenever the percolation function satisfies this "finite derivative condition" at the critical point.

Outline of Proof. By countable additivity, it suffices to show that there are no times at which the origin percolates during $[0, 1]$. We use a first moment argument. We break the time interval $[0,1]$ into m intervals each of length $1/m$. If we fix one of these intervals, the set of edges which are open at *some time* during this interval is i.i.d. with density about $p_c + 1/m$. Hence the probability that the origin percolates with respect to these set of edges is by (11.1) at most $O(1/m)$. It follows that if N_m is the number of intervals where this occurs, then $\mathbb{E}[N_m]$ is at most $O(1)$. It is not hard to check that $N \le \liminf_m N_m$, where N is the cardinality of the set of times during $[0,1]$ at which the origin percolates. Fatou's Lemma now yields that $\mathbb{E}(N) < \infty$ and hence there are at most finitely many exceptional times during $[0,1]$ at which the origin percolates. To go from here to having no exceptional times can either be done by using some rather abstract Markov process theory or by a more hands on approach as was done in (HPS97) and that we refer to for details.   □

Remarks (i). It is known that (11.1) holds for any homogeneous tree (see (Gri99) for the binary tree case) and hence there are no exceptional times of percolation in this case also.
(ii). It is was proved by Kesten and Zhang (KZ87) that (11.1) fails for $\mathbb{Z}^2$ and hence the preceding proof method to show that there are no exceptional times fails. This infinite derivative in this case might suggest that there are in fact exceptional times for critical dynamical percolation on $\mathbb{Z}^2$, an important question left open in (HPS97).

11.3 d=2 and BKS

One of the questions posed in (HPS97) and which earlier had been asked by Itai Benjamini was whether there are exceptional times of percolation for $\mathbb{Z}^2$. The following interesting result in (BKS99), although it does not give this, has a similar flavor. Observe that this is a special case of Theorem 1.23, as pointed out earlier.

Theorem 11.2 (BKS99) *Consider an $R \times R$ box on which we run critical dynamical percolation. Let S_R be the number of times during $[0,1]$ at which the configuration changes from having a percolation crossing to not having one. Then*

$$S_R \to \infty \text{ in probability as } R \to \infty.$$

Noise sensitivity of percolation as well as the above theorem tells us that certain large-scale connectivity properties decorrelate very quickly. This suggests that in some vague sense $\omega_t^{p_c}$ "changes" very quickly as time goes on and hence there might be some chance that an infinite cluster appears because we are given many "chances."

In the next section, we begin our study of exceptional times for $\mathbb{Z}^2$ and the hexagonal lattice.

11.4 The second moment method and the spectrum

In this section, we reduce the question of exceptional times to a "second moment method" computation that in turn reduces to questions concerning the spectral behavior for specific Boolean functions involving percolation. Since $p = 1/2$, our dynamics can be equivalently defined by having each edge or hexagon be re-randomized at rate 1.

The key random variable that one needs to look at is

$$X = X_R := \int_0^1 1_{0 \overset{\omega_t}{\longleftrightarrow} R} \, dt$$

where $0 \overset{\omega_t}{\longleftrightarrow} R$ is of course the event that at time t there is an open path from the origin to distance R away. Note that the above integral is simply the Lebesgue measure of the set of times in $[0,1]$ at which this occurs.

We want to apply the second moment method here. We isolate the easy part of the argument so that readers who are not familiar with this method understand it in a more general context. However, readers should keep in mind that the difficult part is always to prove the needed bound on the second moments which in this case is (11.2).

Proposition 11.3 *If there exists a constant C such that for all R*

$$\mathbb{E}(X_R^2) \le C\mathbb{E}(X_R)^2, \tag{11.2}$$

then a.s. there are exceptional times of percolation.

Proof For any nonnegative random variable Y, the Cauchy–Schwarz inequality applied to $YI_{\{Y>0\}}$ yields

$$\mathbb{P}(Y > 0) \ge \mathbb{E}(Y)^2/\mathbb{E}(Y^2).$$

Hence by (11.2), we have that for all R,

$$\mathbb{P}(X_R > 0) \ge 1/C$$

and hence by countable additivity (as we have a decreasing sequence of events)

$$\mathbb{P}(\cap_R \{X_R > 0\}) \ge 1/C.$$

Had the set of times that a fixed edge is on been a closed set, then the above would have yielded by compactness that there is an exceptional time of percolation with probability at least $1/C$. However, this is not a closed set. On the other hand, this point is very easily fixed by modifying the process so that the times each edge is on is a closed set and observing that a.s. no new times of percolation are introduced by this modification. The details are left to the readers. Once we have an exceptional time with positive probability, ergodicity immediately implies that this occurs a.s. □

The first moment of X_R is, due to Fubini's Theorem, simply the probability of our one-arm event, namely $\alpha_1(R)$. The second moment of X_R is easily seen to be

$$\mathbb{E}(X^2) = \mathbb{E}\left(\int_0^1 \int_0^1 1_{0 \xleftrightarrow{\omega_s} R} 1_{0 \xleftrightarrow{\omega_t} R} \, ds \, dt \right)$$

$$= \int_0^1 \int_0^1 \mathbb{P}(0 \xleftrightarrow{\omega_s} R, 0 \xleftrightarrow{\omega_t} R) \, ds \, dt$$

which is, by time invariance, at most

$$2 \int_0^1 \mathbb{P}(0 \xleftrightarrow{\omega_s} R, 0 \xleftrightarrow{\omega_0} R) \, ds. \tag{11.3}$$

The key observation now, which brings us back to noise sensitivity, is that the integrand $\mathbb{P}(0 \xleftrightarrow{\omega_s} R, 0 \xleftrightarrow{\omega_0} R)$ is precisely $\mathbb{E}[f_R(\omega) f_R(\omega_\epsilon)]$ where f_R is the indicator of the event that there is an open path from the origin to distance R away and $\epsilon = 1 - e^{-s}$ because looking at our process at two different times is exactly looking at a configuration and a noisy version.

What we have seen in this subsection is that proving the existence of exceptional times comes down to proving a second moment estimate and furthermore that the integrand in this second moment estimate concerns noise sensitivity, something for which we have already developed a fair number of tools to handle.

11.5 Proof of existence of exceptional times for the hexagonal lattice via randomized algorithms

In (SS10), exceptional times were shown to exist for the hexagonal lattice; this was the first transitive graph for which such a result was obtained. However, the methods in this paper did not allow the authors to prove that $\mathbb{Z}^2$ had exceptional times.

Theorem 11.4 (SS10)   *For dynamical percolation on the hexagonal lattice* $\mathbb{T}$ *at the critical point* $p_c = 1/2$, *there exist almost surely exceptional times* $t \in [0, \infty)$ *such that* ω_t *has an infinite cluster.*

Proof   As we noted in the previous section, two different times of our model can be viewed as "noising" where the probability that a hexagon is re-randomized within t units of time is $1 - e^{-t}$. Hence, by (4.2), we have that

$$\mathbb{P}[0 \overset{\omega_0}{\longleftrightarrow} R, 0 \overset{\omega_t}{\longleftrightarrow} R] = \mathbb{E}[f_R]^2 + \sum_{\emptyset \neq S \subseteq B(0,R)} \hat{f}_R(S)^2 \exp(-t|S|) \qquad (11.4)$$

where $B(0,R)$ are the set of hexagons involved in the event f_R. We see in this expression that, for small times t, the frequencies contributing in the correlation between $\{0 \overset{\omega_0}{\longleftrightarrow} R\}$ and $\{0 \overset{\omega_t}{\longleftrightarrow} R\}$ are of "small" size $|S| \lesssim 1/t$. Therefore, To detect the existence of exceptional times, one needs to achieve good control on the *lower tail* of the Fourier spectrum of f_R.

The approach of this section is to find an algorithm minimizing the revealment as much as possible and to apply Theorem 8.2. However, there is a difficulty here, as our algorithm might have to look near the origin, in which case it is difficult to keep the revealment small. There are other reasons for a potential problem. If R is very large and t very small, then if one conditions on the event $\{0 \overset{\omega_0}{\longleftrightarrow} R\}$, because few sites are updated, the open path in ω_0 from 0 to distance R will still be preserved in ω_t at least up to some distance $L(t)$ (further away, large scale connections start to decorrelate). In some sense the geometry associated to the event $\{0 \overset{\omega}{\longleftrightarrow} R\}$ is "frozen" on a certain scale between time 0 and time t. Therefore, it is natural to divide our correlation analysis into two scales: the ball of radius $r = r(t)$ and the annulus from $r(t)$ to R. Obviously the "frozen radius" $r = r(t)$ increases as $t \to 0$. We therefore proceed as follows instead. For any r, we have

$$\mathbb{P}[0 \overset{\omega_0}{\longleftrightarrow} R, 0 \overset{\omega_t}{\longleftrightarrow} R] \leq \mathbb{P}[0 \overset{\omega_0}{\longleftrightarrow} r]\mathbb{P}[r \overset{\omega_0}{\longleftrightarrow} R, r \overset{\omega_t}{\longleftrightarrow} R]$$
$$\leq \alpha_1(r)\mathbb{E}[f_{r,R}(\omega_0)f_{r,R}(\omega_t)], \qquad (11.5)$$

where $f_{r,R}$ is the indicator function of the event, denoted by $r \overset{\omega}{\longleftrightarrow} R$, that there is an open path from distance r away to distance R away. Now, as above, we have

$$\mathbb{E}[f_{r,R}(\omega_0)f_{r,R}(\omega_t)] \leq \mathbb{E}[f_{r,R}]^2 + \sum_{k=1}^{\infty} \exp(-tk) \sum_{|S|=k} \hat{f}_{r,R}(S)^2. \qquad (11.6)$$

The Boolean function $f_{r,R}$ somehow avoids the singularity at the origin, and it is possible to find algorithms for this function with small revealments. In any case, letting $\delta = \delta_{r,R}$ be the revealment of $f_{r,R}$, it follows from Theorem 8.2 and the fact that $\sum_k k\exp(-tk) \le O(1)/t^2$ that

$$\mathbb{E}[f_{r,R}(\omega_0)f_{r,R}(\omega_t)] \le \alpha_1(r,R)^2 + O(1)\delta\alpha_1(r,R)/t^2. \tag{11.7}$$

The following proposition gives a bound on δ. We sketch why it is true afterwards.

Proposition 11.5 (SS10) *Let $2 \le r < R$. Then*

$$\delta_{r,R} \le O(1)\alpha_1(r,R)\alpha_2(r). \tag{11.8}$$

Putting together (11.5), (11.7), Proposition 11.5 and using quasi-multiplicativity of α_1 yields

$$\mathbb{P}[0 \overset{\omega_0}{\longleftrightarrow} R, 0 \overset{\omega_t}{\longleftrightarrow} R] \le O(1)\frac{\alpha_1(R)^2}{\alpha_1(r)}\left(1 + \frac{\alpha_2(r)}{t^2}\right).$$

This is true for all r and t. If we choose $r = r(t) = (1/t)^8$ and ignore $o(1)$ terms in the critical exponents (which can easily be handled rigorously), we obtain, using the explicit values for the one- and two-arm critical exponents, that

$$\mathbb{P}[0 \overset{\omega_0}{\longleftrightarrow} R, 0 \overset{\omega_t}{\longleftrightarrow} R] \le O(1)t^{-5/6}\alpha_1(R)^2. \tag{11.9}$$

Now, because $\int_0^1 t^{-5/6}dt < \infty$, by integrating the above correlation bound over the unit interval, one obtains that $\mathbb{E}[X_R^2] \le C\mathbb{E}[X_R]^2$ for some constant C as desired. $\qquad\square$

Outline of proof of Proposition 11.5

We use an algorithm that mimics the one we used for percolation crossings except the present setup is "radial." As in the chordal case, we randomize the starting point of our exploration process by choosing a site uniformly on the "circle" of radius R. Then, we explore the configuration with an exploration path γ directed toward the origin; this means that as in the case of crossings, when the interface encounters an open (resp. closed) site, it turns say to the left (resp. right), the only difference being that when the exploration path closes a loop around the origin, it continues its exploration inside the connected component of the origin. (It is known that this discrete curve converges towards *radial* SLE$_6$ on $\mathbb{T}$, when the mesh goes to 0.) It turns out that the so-defined exploration path gives all the information we need. Indeed, if the exploration path closes a clockwise loop around the

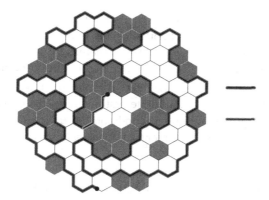

Figure 11.1 Illustration of the algorithm to determine if there is a radial path.

origin, this means that there is a closed circuit around the origin making $f_{r,R}$ equal to 0. On the other hand, if the exploration path does not close any clockwise loop until it reaches radius r, it means that $f_{r,R} = 1$. Hence, we run the exploration path until either it closes a clockwise loop or it reaches radius r. This is our algorithm. See Figure 11.1. Neglecting boundary issues (points near radius r or R), if x is a point at distance u from 0, with $2r < u < R/2$, for x to be examined by the algorithm, it is needed that there is an open path from $2u$ to R and the two-arm event holds in the ball centered at u with radius $u/2$. Hence for $|x| = u$, $\mathbb{P}[x \in J]$ is at most $O(1)\alpha_2(u)\alpha_1(u,R)$. Due to the explicit values of the one- and two-arm exponents, this expression is decreasing in u. Hence, ignoring the boundary, the revealment is at most $O(1)\alpha_2(r)\alpha_1(r,R)$. See (SS10) for more details.   □

We now assume that readers are familiar with the notion of Hausdorff dimension. We let $\mathcal{E} \subseteq [0,\infty]$ denote the (random) set of these exceptional times at which percolation occurs. It is an immediate consequence of Fubini's Theorem that $\mathcal{E}$ has Lebesgue measure 0 and hence we should look at its Hausdorff dimension if we want to measure its "size." The first result is the following.

Theorem 11.6 (SS10) *The Hausdorff dimension of $\mathcal{E}$ is an almost sure constant in* $[1/6, 31/36]$.

It was conjectured there that the dimension of the set of exceptional times is a.s. $31/36$.

Outline of Proof. The fact that the dimension is an almost sure constant follows from easy 0–1 Laws. The lower bounds are obtained by placing a random measure on $\mathcal{E}$ with finite so-called α–energies for any $\alpha < 1/6$ and using a result called Frostman's Theorem. (This is a standard technique once one has good control of the correlation structure.) Basically, the $1/6$ comes from the fact that for any $\alpha < 1/6$, one can multiply the integrand in $\int_0^1 t^{-5/6} dt$ by $(1/t)^\alpha$ and still be integrable. It is the amount of "room to spare" you have. If one could obtain better estimates on the correlations, one could thereby improve the lower bounds on the dimension. The upper bound is obtained via a first moment argument similar to the proof of Theorem 11.1 but now using (2.1).   □

Before moving on to our final method of dealing with the spectrum, let us consider what we might have lost in the above argument. Using the above argument, we optimized things by taking $r(t) = (1/t)^8$. However, at time t compared to time 0, we have noise which is about t. Because we now know the exact noise sensitivity exponent, to obtain decorrelation, the noise level should be at least about the negative 3/4th power of the radius of the region we are looking at. So, events in our annulus should decorrelate if $r(t) >> (1/t)^{4/3}$. This suggests there might be potential for improvement. Note we used an inner radius that is 6 times larger than potentially necessary ($8 = 6 \times 4/3$). This 6 is the same 6 by which the result in Theorem 8.5 differed by the true exponent ($3/4 = 6 \times 1/8$) and the same 6 explaining the gap in Theorem 11.6 $(1 - 1/6) = 6 \times (1 - 31/36)$. This last difference is also seen by comparing the exponents in (11.9) and the last term in (11.10) that follows.

11.6 Proof of existence of exceptional times via the geometric approach of the spectrum

Recall that our third approach for proving the noise sensitivity of percolation crossings was based on a geometrical analysis of the spectrum, viewing the spectrum as a random set. This approach yielded the exact noise sensitivity exponent for percolation crossings for the hexagonal lattice. This approach can also be used here, as we now explain. Two big advantages of this approach are that it succeeded in proving the existence of exceptional times for percolation crossings on $\mathbb{Z}^2$, something that (SS10) was not able to do, as well as obtaining the exact Hausdorff dimension for the set of exceptional times, namely the upper bound of $31/36$ in the previous result.

Theorem 11.7 (GPS10)   *For the triangular lattice, the Hausdorff dimension of $\mathcal{E}$ is almost surely $31/36$.*

Proof As explained in the previous section, it suffices to lower the 5/6 in (11.9) to 5/36. (Note that (11.9) was really only obtained for numbers strictly larger than 5/6, with the $O(1)$ depending on this number; the same will be true for the 5/36.)

Let $s(r)$ be the inverse of the map $r \to r^2\alpha_4(r) \sim r^{3/4}$. So more or less, $s(r) := r^{4/3}$. Using Theorem 10.22, we obtain the following:

$$\mathbb{E}[f_R(\omega_0)f_R(\omega_t)] = \sum_S \exp(-t|S|)\hat{f}_R(S)^2$$

$$= \sum_{k=1}^{\infty} \sum_{S\,:\,|S|\in[(k-1)/t,k/t)} \exp(-t|S|)\hat{f}_R(S)^2$$

$$\leq \sum_{k=1}^{\infty} \exp(-k)\hat{\mathbb{Q}}[|\mathscr{S}_{f_R}| < k/t]$$

$$\leq O(1) \sum_{k=1}^{\infty} \exp(-k)\frac{\alpha_1(R)^2}{\alpha_1(s(k/t))}$$

$$\leq O(1)\alpha_1(R)^2 \sum_{k=1}^{\infty} \exp(-k)(\frac{k}{t})^{4/3\times 5/48}$$

$$\leq O(1)\alpha_1(R)^2(\frac{1}{t})^{5/36}. \tag{11.10}$$

This completes the proof. (Of course, there are $o(1)$ terms in these exponents which we are ignoring.) □

We have done a lot of the work for proving that there are exceptional times also on $\mathbb{Z}^2$.

Theorem 11.8 (GPS10) *For dynamical percolation on $\mathbb{Z}^2$ at the critical point $p_c = 1/2$, there exist almost surely exceptional times $t \in [0,\infty)$ such that ω_t has an infinite cluster.*

Proof $s(r)$ is defined as it was before but now we cannot say that $s(r)$ is about $r^{4/3}$. However, we can say that for some fixed $\delta > 0$, we have that for all r,

$$s(r) \geq r^{\delta} \tag{11.11}$$

From the previous proof, we still have

$$\frac{\mathbb{E}[f_R(\omega_0)f_R(\omega_t)]}{\alpha_1(R)^2} \leq O(1) \sum_{k=1}^{\infty} \exp(-k)\frac{1}{\alpha_1(s(k/t))}. \tag{11.12}$$

Exactly as in the proof of Theorem 11.4, we need to show that the right hand side is integrable near 0 in order to carry out the second moment argument.

Quasi-multiplicativity can be used to show that

$$\alpha_1(s(1/t)) \leq k^{O(1)} \alpha_1(s(k/t)). \tag{11.13}$$

(Note that if things behaved exactly as power laws, this would be clear.)

Therefore the above sum is at most

$$O(1) \sum_{k=1}^{\infty} \exp(-k) \frac{k^{O(1)}}{\alpha_1(s(1/t))} \leq O(1) \frac{1}{\alpha_1(s(1/t))} \tag{11.14}$$

V. Beffara has shown (see the appendix in (GPS 10) for a proof) that there exists $\epsilon_0 > 0$ such that for all r,

$$\alpha_1(r)\alpha_4(r) \geq r^{\epsilon_0 - 2}. \tag{11.15}$$

Note that Theorem 6.4 and (6.7) tell us that the left hand side is larger than $\Omega(1)r^{-2}$. The above tells us that we get an (important) extra power of r in (6.7).

It follows that

$$\frac{1}{\alpha_1(s(1/t))} \leq \alpha_4(s(1/t))s(1/t)^{2-\epsilon_0} = (1/t)s(1/t)^{-\epsilon_0}. \tag{11.16}$$

Equation (11.11) tells us that the last factor is at most t^η for some $\eta > 0$ and hence the relevant integral converges as desired. The rest of the argument is the same.   □

One can also consider exceptional times for other events, such as for example times at which there is an infinite cluster in the upper half-plane or times at which there are two infinite clusters in the whole plane, and consider the corresponding Hausdorff dimension. A number of results of this type, which are not sharp, are given in (SS 10) while various sharp results are given in (GPS 10).

An example of such a result is the following.

Theorem 11.9 (GPS 10) *For dynamical percolation on* $\mathbb{T}$, *there exist times at which both an infinite black cluster and an infinite white cluster exist. Moreover, the Hausdorff dimension of this set of times is at least* $1/9$ *a.s.*

We finally mention that at the time that (HPS 97) was written, there were discussions concerning the question of a possible relationship between the so-called incipient infinite cluster constructed by H. Kesten and exceptional times for 2-d dynamical percolation should the latter exist. In (HPS 13), such a relation is obtained by placing a local time measure on the set of exceptional times.

11.7 Exercises

11.1 Prove that on any graph below criticality, there are no times at which there is an infinite cluster while above criticality, there is an infinite cluster at all times.

11.2 Consider critical dynamical percolation on a general graph satisfying $\theta(p_c) = 0$. Show that a.s. $\{t : \omega_t$ percolates $\}$ has Lebesgue measure 0.

11.3 (Somewhat hard). A *spherically symmetric* tree is one where all vertices at a given level have the same number of children, although this number may depend on the given level. Let T_n be the number of vertices at the nth level. Show that there is percolation at p if

$$\sum_n \frac{1}{p^{-n}T_n} < \infty$$

Hint: Let X_n be the number of vertices in the nth level that are connected to the root. Apply the second moment method to the sequence of X_n's.

The convergence of the sum is also necessary for percolation but this is harder and you are not asked to show this. This theorem is due to Russell Lyons.

11.4 Show that if T_n is $n^2 2^n$ up to multiplicative constants, then the critical value of the graph is $1/2$ and we percolate at the critical value. (This yields a graph that percolates at the critical value.)

11.5 (Quite a bit harder). Consider dynamical percolation on a spherically symmetric tree. Show that there for the parameter p, there are exceptional times at which percolation occurs if

$$\sum_n \frac{1}{np^{-n}T_n} < \infty.$$

Hint: Find an appropriate random variable X_n to which the second moment method can be applied.

11.6 Find a spherically symmetric tree that does not percolate at criticality but for which there are exceptional times at which percolation occurs.

For the connoisseur

In this chapter, we briefly discuss a number of miscellaneous topics related to this subject which we find interesting. Most are not directly connected to percolation, which has been the major focus for us up to this point. Most of the sections in this chapter are independent of each other.

12.1 How noise sensitive can monotone functions be?

Crossing events for percolation on the triangular lattice yield for us monotone events all of whose spectrum is near $n^{3/8}$, where n is the number of bits (note that the 3/4 is now 3/8 because we are indexing by the number of bits rather than the side length of the box).

On the other hand, Proposition 4.6 (together with Proposition 4.5) tells us that the expected size of the spectral sample for monotone functions on n bits taking values in $\{\pm 1\}$ is at most $n^{1/2}$. Markov's inequality then implies that for such monotone functions on n bits, the probability that the spectral sample lies above $n^{1/2+\delta}$ goes to 0, with n for any fixed positive δ.

In (BKS99), a question was asked that, although different than stated below, we believe was intended to be the following. Does there exist $\beta < 1/2$ such that for every sequence of monotone Boolean functions on n bits taking values in $\{\pm 1\}$, the probability that the spectral sample exceeds n^β goes to 0 with n? If true, Iterated 3-Majority shows that such a β must be at least $1 - \log 2/\log 3$. If true, this would be equivalent to the existence of a $\delta_0 > 0$ so that for any sequence $\{f_n\}$ of monotone Boolean functions on n bits taking values in $\{\pm 1\}$,

$$\lim_{n\to\infty} \mathbb{P}[f_n(\omega) \neq f_n(\omega_{1/n^{1/2-\delta_0}})] = 0.$$

In (MO03), it was shown that the answer is no in a strong sense.

Theorem 12.1 (MO03) *For each $\delta > 0$, there exists a sequence $\{f_n\}$ of monotone Boolean functions on m_n bits (with m_n going to ∞) mapping into*

$\{-1, 1\}$ *whose spectral samples* $\{\mathscr{S}_n\}$ *satisfy*

$$\lim_{n \to \infty} \hat{\mathbb{P}}[|\mathscr{S}_n| \le m_n^{1/2-\delta}] = 0.$$

Owing to the relationship between spectrum and noise, this is equivalent to saying that there is a sequence $\{f_n\}$ of monotone Boolean functions on m_n bits with

$$\lim_{n \to \infty} \mathbb{E}(f_n(\omega) f_n(\omega_{1/m_n^{1/2-\delta}})) = 0.$$

We do not give the proof of this but will present the example. Given $\delta > 0$, one chooses $k = k(\delta)$ odd and sufficiently large and then considers Iterated k-Majority, which is the obvious generalization of Iterated 3-Majority. The proof is based on the analysis of a certain recursion.

However, here we give a proof of a weaker result that still answers the question from (BKS99) and is based on more elementary considerations such as those given in the proof of Theorem 9.8.

Theorem 12.2 *For each $\delta > 0$, there exists a sequence $\{f_n\}$ of monotone Boolean functions $\{f_n\}$ on m_n bits (with m_n going to ∞) mapping into $\{-1, 1\}$ whose spectral samples $\{\mathscr{S}_n\}$ satisfy*

$$\liminf_{n \to \infty} \hat{\mathbb{P}}[|\mathscr{S}_n| \ge m_n^{1/2-\delta}] > 0.$$

Owing to the relationship between spectrum and noise, this is equivalent to saying that there are monotone Boolean functions with

$$\liminf_{n \to \infty} \mathbb{P}[f_n(\omega) \ne f_n(\omega_{1/m_n^{1/2-\delta}})] > 0,$$

which is to say that $f_n(\omega)$ and $f_n(\omega_{1/m_n^{1/2-\delta}})$ are "at least partially independent."

Proof Fix $\delta > 0$. For k odd, letting a_k denote the probability that simple random walk is back at the origin at time $k - 1$, it is known that a_k behaves like $c/k^{1/2}$ for large k and hence that $ka_k \ge 2k^{1/2-\delta}$ for large k. Choose such an odd k. We now consider the variant of the Iterated 3-Majority function already mentioned where 3 is replaced by k. We thereby obtain a sequence of Boolean functions $\{f_n\}$ with f_n having $m_n = k^n$ bits where n represents the number of levels in the tree. Let $\mathscr{S}_n$ and $\mathcal{P}_n$ denote respectively the spectral measure and the pivotal set for f_n.

It is easy to check that $\mathbb{E}|\mathcal{P}_n| = (ka_k)^n$ and hence by Proposition 9.4 that $\hat{\mathbb{E}}[|\mathscr{S}_n|] = (ka_k)^n$. This gives $\hat{\mathbb{E}}[|\mathscr{S}_n|] \ge 2m_n^{1/2-\delta}$. for all n.

Next, a relatively easy computation, left to the readers, shows that there is a constant $C_k < \infty$ so that for all n

$$\mathbb{E}|\mathcal{P}_n|^2 \leq C_k \mathbb{E}|\mathcal{P}_n|^2. \tag{12.1}$$

Corollary 9.7 then yields (as in the proof of Theorem 9.8) that

$$\hat{\mathbb{E}}|\mathcal{S}_n|^2 \leq C_k \hat{\mathbb{E}}|\mathcal{S}_n|^2. \tag{12.2}$$

The Paley–Zygmund inequality now yields that for all n

$$\hat{\mathbb{P}}[|\mathcal{S}_n| \geq m_n^{1/2-\delta}] \geq \hat{\mathbb{P}}[|\mathcal{S}_n| \geq \hat{\mathbb{E}}[|\mathcal{S}_n|]/2] \geq 1/(4C_k).$$

$\square$

We emphasize the difference between Theorems 12.1 and 12.2. The former says that we can get "all" the mass near $n^{1/2}$ while the latter says we can get "some nontrivial portion" of the mass near $n^{1/2}$. Equivalently, the first says we can "completely decorrelate" at noise levels slightly higher than $(1/n)^{1/2}$ while the latter says we can "at least partially decorrelate" at this noise level. An idea of proving Theorem 12.1 along the lines of Theorem 12.2 is to use the iterated majority functions with k also growing. This idea would work if the constants C_k in the proof of Theorem 12.2 could be taken to go to 1 with k. Unfortunately, they necessarily tend to ∞ and so this method fails.

Interesting, Theorem 12.2 can be strengthened by not only having nontrivial spectral measure "near" $n^{1/2}$ but actually at $n^{1/2}$ itself. The following theorem is proved in (MO03); the construction is based on a construction of Talagrand and is a random construction. The proof is not given.

Theorem 12.3 (MO03)   *There exists a sequence $\{f_n\}$ of monotone Boolean functions on n bits mapping into $\{-1,1\}$ such that*

$$\inf_n \mathbb{P}[f_n(\omega) \neq f_n(\omega_{1/n^{1/2}})] > 0. \tag{12.3}$$

Remarks (1) The claim of this theorem is equivalent to that there exists $c > 0$ so that for all n,

$$\hat{\mathbb{P}}[|\mathcal{S}_n| \geq cn^{1/2}] \geq c.$$

(2) The exponent of $1/2$ cannot be improved.
(3) The asympototic "partial independence" of $f_n(\omega)$ and $f_n(\omega_{1/n^{1/2}})$ cannot be improved to full independence in the sense that $E(f_n(\omega)f_n(\omega_{1/n^{1/2}}))$ approaches 0.
(4) In fact, such a sequence cannot even be noise sensitive because (1) and Proposition 9.4 imply that the total influence is $\Omega(n^{1/2})$ and from this, the

Cauchy–Schwarz inequality yields that the sum of the squared influences is $\Omega(1)$.

We make a few comments concerning the example in Theorem 12.3. It is constructed by using a random conjunctive normal form (CNF) formula. One takes the "AND" of $2^{n^{1/2}}$ independent "OR" clauses each of length $n^{1/2}$ where each element is chosen uniformly and independently from $[n]$. One then shows that with an $\Omega(1)$ probability, the function satisfies the desired properties. Talagrand (Tal96) originally constructed this example to find a sequence of Boolean functions satisfying

$$\mathbb{P}[|\mathcal{P}_n| \geq cn^{1/2}] \geq c.$$

Given the relationship between $\mathcal{P}_n$ and $\mathcal{S}_n$, this is similar to (1) but of course neither implies the other.

12.2 Noise sensitivity and the pivotal set

Recall that for Boolean functions f mapping into $\{\pm 1\}$, $\mathcal{P}_f$ and $\mathcal{S}_f$ have identical first and second marginals; this yields the fact that the expected size of these two random sets are equal. In this section, we show nonetheless that these random sets can be very different and so in particular, knowing certain properties of the pivotal set does not yield information concerning noise sensitivity questions.

This first example gives a simple example that is noise sensitive but where the pivotal set is empty with probability going to 1; these two properties are immediate.

Example 12.4 Let f be the Boolean function, which is 1 if and only if $\lfloor N(\omega)/\log n \rfloor$ is even where $N(\omega)$ is the number of 1's.

More interestingly, we have the following proposition giving a monotone such example.

Proposition 12.5 *There exists a sequence* $\{f_n\}$ *of monotone Boolean functions which is noise sensitive but for which*

$$\lim_{n \to \infty} \mathbb{P}(\mathcal{P}_{f_n} = \emptyset) = 1.$$

Proof Consider percolation on an $m \times m$ box in the triangular lattice. Replace each bit by an odd number $k = k_m$ of bits so that overall we have km^2 bits. f_m is taken to be a composition of Majority on k_m bits with percolation; in other words each hexagon in our grid is open if the majority of the k_m

bits corresponding to it is in state 1 and then we see if there is a percolation crossing.

It is easy to show that no matter how k_m is chosen, the sequence $\{f_m\}$ is noise sensitive; this uses the fact that Dictator is the most stable Boolean function with mean 0.

Next, with probability going to 1, there are at most $m^{.76}$ pivotal sites for the crossing event. But each site has probability only about $1/k^{1/2}$ to have a pivotal bit. Therefore, if $m^{.76}/k^{1/2} \to 0$ when $m \to \infty$, then we would have what we want. So, simply let $k_m = m^2$ and we are done.   □

Although the above shows that noise sensitivity can occur even if $\mathcal{P}_n$ is typically empty, there are still restrictions. Certainly, $\mathbb{E}(|\mathcal{P}_n|)$ must tend to infinity. However, much stronger, Talagrand (see (Tal97)) showed that any nondegenerate sequence of monotone functions that is noise sensitive necessarily satisfies that $\mathbb{E}|\mathcal{P}_n|^{1/2}$ tends to infinity. Modifications of the example in the proof of Proposition 12.5 show that the exponent of $1/2$ in Talagrand's result is sharp.

The next two theorems in this section, Theorems 12.6 and 12.8, were the outcome of conversations with Ryan O'Donnell and Devdatt Dubhashi. The first result shows that we can have order n pivotal bits always but nonetheless not have noise sensitivity.

Theorem 12.6   *There exists a sequence m_n going to ∞, a sequence $\{f_n\}$ of mean $1/2$ Boolean functions on m_n bits mapping into $\{0,1\}$ and $c > 0$ so that the sequence is not noise sensitive but for all n, $\mathbb{P}(|\mathcal{P}_{f_n}| \geq cm_n) = 1$.*

We first need the following

Lemma 12.7   *For every $p \in [1/2, 1)$, there exists $c > 0$ and a sequence $\{f_n\}$ of Boolean functions on m_n variables with mean at least p such that for all n, $\mathbb{P}(|\mathcal{P}_{f_n}| \geq cm_n) = 1$.*

Proof   The key step is to show for each such p, there is a Boolean function f with mean at least p such that there is always at least one pivotal bit. Once we have this, if this function has k bits, we can define, for any n, a function f_n on nk bits by composing f with the Parity function on n bits. That is, we replace each bit of f by a Parity function on n bits. This yields what we want with $m_n = nk$ and c being $1/k$.

To obtain a function with mean at least p that always has at least one pivotal bit, we do a random construction. Let $p' = (p + 1)/2$ and for each k, we construct a random Boolean function on k bits by assigning to each of the 2^k input strings, the value 1 with probability p' and the value 0 with probability $1 - p'$, independently for different strings. By standard large deviations theory, the probability that the random Boolean function does not have mean at least p is at most $e^{-c_1 2^k}$ for some constant c_1 depending on p. We now show using the Lovász local lemma that for large k, the probability that the random Boolean function always has at least one pivotal is at least $e^{-c_2(p'2)^k}$ for some constant c_2 depending on p. Together, this gives us with positive probability a function of the desired form.

To apply the Lovász local lemma, we let for each $\omega \in \Omega_k$, let E_ω be the event that the (random) function has no pivotals at ω. It is clear that for each ω, $P(E_\omega) \le 2(p')^{k+1}$ and that E_ω is jointly independent of all the events $E_{\omega'}$ for ω' outside of the 2-neighborhood of ω. Lemma 5.1.1 in (AS00) (which is a general version of the Lovász local lemma with an explicit lower bound) yields the claimed lower bound $e^{-c_2(p'2)^k}$ above as desired; one can take x_i to be $4(p')^k$ in this lemma. □

Proof of Theorem 12.6. Choose a sequence $\{f_n\}$ of Boolean functions as in Lemma 12.7 with $p = 0.8$ and let g_n be f_n XORed with a single bit. (The number of bits for g_n is $m_n + 1$.) Then g_n has mean $1/2$ and the number of pivotals bits of g_n is 1 plus the number of pivotals bits for f_n (when looking at its domain); hence the proportion of pivotal bits is always at least c where c comes from Lemma 12.7.

We now show the sequence $\{g_n\}$ is not noise sensitive. Fix $\epsilon < 0.05$. Then $\mathbb{P}(g(\omega) = g(\omega_\epsilon))$ is at least the probability that the last bit does not re-randomize and that f at both ω with the last bit removed and at ω_ϵ with the last bit removed is 1. The probability of this occuring is at least $(1 - \epsilon)(1 - 2(.2)) \ge .51$. Hence noise sensitivity fails. □

The next result shows that we can have that the number of pivotal bits is always any power of n less than 1 but still have noise stability.

Theorem 12.8 *For each $\epsilon > 0$, there exists a sequence $\{f_n\}$ of mean $1/2$ Boolean functions on m_n bits mapping into $\{0, 1\}$ so that the sequence is noise stable but for all n, $\mathbb{P}(|\mathcal{P}_{f_n}| \ge m_n^{1-\epsilon}) = 1$.*

Proof Using Lemma 12.7, for each k, choose a sequence $f_n^{(k)}$ of Boolean functions all of whose means are at least $1 - 1/k$ and such that for each n, the number of pivotals has density at least $c_k > 0$ with probability 1. Because for any fixed $c_k > 0$, $c_k n \ge n^{1-\epsilon}$ for all large n, the sequence of Boolean functions constructed by first starting with $f_n^{(1)}$, then at a sufficiently large time switch to the sequence $f_n^{(2)}$, then at a sufficiently large time switch to the sequence $f_n^{(3)}$, etc. would yield a sequence satisfying for all n, $\mathbb{P}(|\mathcal{P}_{f_n}| \ge m_n^{1-\epsilon}) = 1$. If we XOR these with a single bit as in Theorem 12.8, this would give us mean $1/2$ functions with as many pivotals while the argument in Theorem 12.8 immediate shows that the sequence is noise stable.   □

We end this section with an example due to Oded Schramm of a nondegenerate sequence of *monotone* functions that is noise stable but such that when the event occurs, there will typically be many pivotals.

Theorem 12.9   *(O. Schramm). There exists a sequence* $\{A_n\}$ *of monotone functions whose variances are* $\Omega(1)$ *that is noise stable but such that for some* b_n *going to* ∞, *we have that*

$$\lim_{n\to\infty} \mathbb{P}(|\mathcal{P}_{A_n}| \geq b_n | A_n^c) = 1.$$

Proof   Choose a sequence of integers a_k so that $\sum_k (1/2)^{a_k} < \infty$ but $\sum_k a_k (1/2)^{a_k} = \infty$. Let G_n be a graph that is a disjoint union of n paths, where the lengths of the paths are $a_1, a_2, \ldots, a_n$. Perform percolation with $p = 1/2$ on the edges and let A_n be the event that at least one of the n paths has all of its edges on. Since $\sum_k (1/2)^{a_k} < \infty$, the expected number of paths with all its edges open is $O(1)$ from which it is easy to conclude that the variances are $\Omega(1)$. Next, it is an easy exercise to show, using the fact that the A_n's are decreasing in n, that this sequence is noise stable. Finally, the fact that $\sum_k a_k (1/2)^{a_k} = \infty$ easily implies that the expected number of paths with exactly one closed edge goes to ∞ with n. Using independence, we obtain the fact that the number of such paths goes to ∞ in probability which easily gives the last claim. (On the other hand, on the event A_n, the number of pivotals is bounded in probability.)   □

12.3 Influences for monotone systems and applications to phase transitions

The results of KKL can be extended to certain nonproduct measures. See (GG06) for details and extensions of the result described in this section. Consider a sequence of $\{0, 1\}$ random variables $X_1, \ldots, X_n$ not necessarily independent. We can think of these as giving us a probability measure μ on $\{0, 1\}^n$. It turns out to be natural to restrict to a certain class of measures, called monotone measures and for simplicity, we will also assume that our measures have full support.

Definition 12.10 $X_1, \ldots, X_n$ is **monotone** if for each i and for each pair of realizations η and δ for the other bits satisfying $\eta \leq \delta$, we have

$$\mathbb{P}(X_i = 1 | \{X_j\}_{j\neq i} = \eta) \leq \mathbb{P}(X_i = 1 | \{X_j\}_{j\neq i} = \delta).$$

How should one now define the influence of a variable on an *increasing* event A in this more general context? If one defines it to be $\mathbb{P}(i$ is pivotal for $A)$, then the analog of Theorem 1.14 fails for the Majority event A_n on n bits even for the reasonably nice monotone measure which is a convex combination of two product measures, one with density $1/3$ and the other with density $2/3$. A_n has probability $1/2$ but it is easy to see that for each i, $\mathbb{P}(i$ is pivotal for $A)$ is exponentially small in n.

The proper definition for monotone variables and an increasing event A is to consider

$$\mathbf{I}_i(A) := \mathbb{E}(A|X_i = 1) - \mathbb{E}(A|X_i = 0)$$

which is easily seen to be equivalent to the usual definition of influence in the i.i.d. case for increasing events. With this definition of influence, we can state the main result in (GG06) which is an extension of Theorems 1.14 and 3.3.

Theorem 12.11 (GG06) *There exists a universal $c > 0$ such that if μ is a monotone measure on $\{0,1\}^n$ with full support and A is an increasing event, then there exists some i such that*

$$\mathbf{I}_i(A) \geq c \min\{\mu(A), 1 - \mu(A)\}(\log n)/n.$$

It turns out that one can also prove a version of Theorems 1.16 and 3.4 in this monotone context. We do not present this here but rather simply refer to (GG11).

Theorem 12.11 is one of the ingredients needed in the proof of the following result which proves the conjectured critical values for the random cluster models in two dimensions; the most important ingredient is a general version of RSW for these models. The random cluster model is a dependent percolation model which is a central object in statistical mechanics. The definition of the model can be found in (BDC12).

Theorem 12.12 (BDC12) *The critical value for the random cluster model with parameter q on the square lattice is*

$$\sqrt{q}/(1 + \sqrt{q}).$$

12.4 The majority is stablest theorem

In this section, we describe an important result concerning stability of the Majority functions $\mathbf{MAJ}_n$; this section demonstrates once again the central role played by these functions. Despite the fact that the results described here hold in greater generality, throughout this section, we stick to Boolean functions mapping into $\{\pm 1\}$ and having mean 0.

For mean 0 Boolean functions, we know that $\mathbb{E}[f(\omega)f(\omega_\epsilon)]$ is for all $\epsilon > 0$ uniquely maximized at the Dictator function $\mathbf{DICT}_n$ (and its equivalent forms). However, for such functions, there is a variable with very large influence and hence such functions are not reasonable to use as a voting scheme. There was a conjecture that if we consider only functions with small influence, then the majority functions $\mathbf{MAJ}_n$ should be the most stable. This conjecture turned out to be true and is the following.

Theorem 12.13 (MOO10)   *For all $\epsilon, \delta > 0$ there exists $\tau > 0$ such that for any mean 0 Boolean function f whose maximum influence is at most τ, we have*

$$\mathbb{E}[f(\omega)f(\omega_\epsilon)] \le \frac{2}{\pi} \arcsin(1 - \epsilon) + \delta.$$

Remark It is the case (as seen in Exercise 4.13) that for each ϵ,

$$\lim_{n \to \infty} \mathbb{E}[\mathbf{MAJ}_n(\omega)\mathbf{MAJ}_n(\omega_\epsilon)] = \frac{2}{\pi} \arcsin(1 - \epsilon);$$

as this is the expression on the right-hand side above, this explains why this result is called *Majority is Stablest*.

We do not prove this result but indicate some of the ingredients in the proof as well as give its motivation.

Ingredients in the proof

The Berry–Esseen central limit theorem stays that if $X_1, \ldots, X_n$ are i.i.d. mean 0, variance 1 random variables with a bounded third moment and $\sum_{i=1}^n c_i^2 = 1$ with the c_i's "small," then $\sum_{i=1}^n c_i X_i$ has approximately a normal distribution (with an explicit bound between the two distributions). The first key step is to generalize this to obtain an invariance principle for multilinear polynomials. Here the limit will no longer be Gaussian but one will obtain a polynomial of independent Gaussians. Here is a first version of the results obtained in (MOO10).

Theorem 12.14 (MOO10)   *Let $X_1, \ldots, X_n$ be i.i.d. mean 0, variance 1 random variables with $\mathbb{E}(|X_1|^3) \le \beta$.*
Consider $Q(X_1, \ldots, X_n) = \sum_{S \subseteq \{1, \ldots, n\}} c_S \prod_{i \in S} X_i$ Assume that (i) $\sum_{S \ne \emptyset} c_S^2 = 1$, (ii) for each i, $\sum_{S : i \in S} c_S^2 \le \tau$ and that (iii) $c_S = 0$ for all S with $|S| > d$. Then if $G_1, \ldots, G_n$ are i.i.d. standard Gaussians, we have that

$$\sup_t |\mathbb{P}(Q(X_1, \ldots, X_n) \le t) - \mathbb{P}(Q(G_1, \ldots, G_n) \le t)| \le O(d\beta^{1/3} \tau^{1/8d}).$$

Remarks (1) All the terms in $Q(X_1, \ldots, X_n)$ are orthogonal and so $\sum_{S \neq \emptyset} c_S^2 = 1$ is simply a normalization so that our expression has variance 1.

(2) If the X_i's were $\{\pm 1\}$ valued and Q a Boolean function, then $\sum_{S : i \in S} c_S^2$ would just be the influence of the ith bit and so assumption (ii) then simply says that all the influences are small.

(3) Assumption (iii) is that Q is a *small degree* multilinear polynomial.

One of the key steps is to use various strengthenings of Theorem 12.14 in order to prove Theorem 12.13 by reducing it to the analogous statement for Gaussian random variables. The case of Gaussian variables was proved by C. Borell; see (Bor85).

Motivation for the theorem

The motivation for proving the Majority is Stablest Conjecture came in fact from approximation algorithms in theoretical computer science. It was shown in (KKMO07) that under a certain strengthening of the $P \neq NP$ hypothesis (technically called the unique games conjecture), the Majority is Stablest Conjecture implies that it is computationally hard to approximate the maximum cut in graphs to within a certain factor better than that which matches the approximation factor achieved by the so-called Goemans and Williamson algorithm.

12.5 k-SAT satisfiability and sharp thresholds

The k-SAT problem is a celebrated problem in computer science. It can be briefly described as follows. Let us fix an integer $k \geq 2$. We will consider a certain class of Boolean functions on the hypercube $\Omega_n := \{-1, 1\}^n$ called k-CNF functions (CNF stands for Conjunctive Normal Form). These Boolean functions are conjunctions (corresponding to an AND operator denoted by $\wedge$) of arbitrary many k-clauses where a k-clause is defined as follows.

Definition 12.15 A **k-clause** on n Boolean variable $(x_1, \ldots, x_n) \in \{-1, 1\}^n$ is a disjunction (i.e., an OR operate denoted by $\vee$) on k variables possibly after taking their negation. For example if $n = 8$ and $k = 3$ the Boolean function

$$x_1 \vee \bar{x}_3 \vee x_7,$$

is a 3-clause.

As such, if $n = 8$, the following Boolean function

$$(x_1 \vee \bar{x}_2 \vee x_7) \wedge (x_2 \vee \bar{x}_3 \vee x_7)$$

is a 3-CNF function.

Definition 12.16 (Satisfiability)

We say that a Boolean function $f : \{-1, 1\}^n \to \{0, 1\}$ is **satisfiable** if one can find at least one configuration $\omega \in \{-1, 1\}^n$ such that $f(\omega) = 1$.

In the **k-SAT Problem**, one is given a k-CNF Boolean function f on n-variables and the problem is to determine whether the function f is satisfiable or not. If $k = 2$, there are polynomially fast algorithms to answer the problem. But for $k \geq 3$, it turns out to be computationally much harder. In fact, the case $k \geq 3$ is one of the first combinatorial problems which has been proved to be in the NP-class. It is easy to check that the above example of a 3-CNF function is satisfiable by choosing $x_1 = x_7 = 1$. An example of a nonsatisfiable 2-CNF is the following.

$$(x_1 \vee x_2) \wedge (\bar{x}_1 \vee \bar{x}_2) \wedge (x_1 \vee \bar{x}_2) \wedge (\bar{x}_1 \vee x_2) \tag{12.4}$$

We now introduce some probability and consider random k-CNF functions in the following sense.

Definition 12.17 (Random k-CNF functions)   Let $1 \leq k \leq n$ be fixed. Let $N = 2^k \binom{n}{k}$ be the number of possible k-clauses. For any fixed parameter $1 \leq M \leq N$, we sample a random k-CNF function by choosing M k-clauses uniformly among the N possible ones and take their conjunction. Similarly as in the study of sharp thresholds (Chapter 3), one can consider the probability that such a random k-CNF function is satisfiable. As in the Erdős–Renyi random graphs $\mathcal{G}(n, p)$, the interesting regime occurs when M is of order cn for some fixed parameter $c \in (0, \infty)$. We denote by $\phi_k(c) = \phi_{k,n}(c)$ the probability that a random k-CNF function made of cn uniformly chosen k-clauses is satisfiable.

It is easy to see that the function $\phi_{k,n}$ is decreasing in c (when c gets larger, there are more clauses that need to be satisfied). Following the notations from (AP04), let us introduce for any $k \geq 2$, the quantities:

$$c_k := \sup\{c \geq 0 : \lim_{n \to \infty} \phi_{k,n}(c) = 1\} \tag{12.5}$$

$$c_k^* := \inf\{c > 0 : \lim_{n \to \infty} \phi_{k,n}(c) = 0\} \tag{12.6}$$

It has been proved (see (AP04) for references) that in the case $k = 2$, one has $c_2 = c_2^* = 1$. The proof in this case relies deeply on a connection with the above Erdős–Renyi graphs. For $k \geq 3$, the situation is still widely open. The main conjecture may be stated as follows.

Conjecture 12.1 *For any $k \geq 3$, one has $0 < c_k = c_k^* < \infty$. In particular, a very interesting* **phase transition** *occurs around the value of c_k.*

These types of phase transitions are usually of high interest to computer scientists because they usually suggest the existence of a threshold between an "approximable phase" (where one may predict in polynomial time with high probability whether the random k-CNF is satisfiable or not) and an inapproximability phase, where even approximability approaches fall in the NP-class. (See (O'D14) for more details and references).

Let us point out that this phase transition even attracted the attention of theoretical physicists who studied this phase transition similarly as they studied the so-called *spin glass* models, by using the *cavity method*. See, for example, (Mez03; MMZ06).

We end this section with the following striking theorem by Ehud Friedgut.

Theorem 12.18 (Friedgut: sharp threshold in the k-SAT problem, (Fri99)) *For any $k \geq 3$, there exists a sequence of functions $(c_k(n))_{n \geq 1}$ bounded away form 0 and ∞ such that for any $\epsilon > 0$,*

$$\lim_{n \to \infty} \phi_{k,n}(c_k(n) - \epsilon) = 1$$

$$\lim_{n \to \infty} \phi_{k,n}(c_k(n) + \epsilon) = 0$$

In particular, this proves a **sharp threshold** *phenomenon.*

Let us briefly explain what the strategy of proof in (Fri99) is. By duality, instead of considering a random k-CNF function f, one can consider its negation $\bar{f}$ which is nothing but a disjunction of M uniformly chosen conjunctions of k-variables. These functions are called k-DNF functions, where DNF stands for Disjunctive Normal Form. Below is an example of such a function with $k = 3$:

$$(x_1 \wedge \bar{x}_2 \wedge x_7) \vee (x_2 \wedge \bar{x}_3 \wedge x_7) \vee (x_5 \wedge x_6 \wedge \bar{x}_8)$$

In this new setting, it is easy to see that the k-SAT problem is translated into the following question: what is the probability that a random k-DNF function is such that it is satisfied for ALL inputs $\omega \in \{-1, 1\}^n$. This probability is now an increasing function of the parameter c such that $M = cn$. One can code this problem using a very useful generalization of graphs: hypergraphs as defined below.

Definition 12.19 An **hypergraph** H is a pair $H = (V, E)$, where V denotes the set of vertices of H and E is any subset of $\mathcal{P}(V) \setminus \{\emptyset\}$. The elements of E are called **hyperedges**. As such, vertices are no longer connected together simply by edges, which in this setting would correspond to pairs of vertices $\{u, v\} \subset V$ but by more general subsets $K \subset V$.

For any $k \geq 1$, a k-**uniform hypergraph** is a hypergraph for which each hyperedge $e \in E$ has cardinality k. For example, a 2-uniform hypergraph corresponds to a (standard) graph.

The k-SAT problem is naturally encoded by a slight generalization of hypergraphs, where $V = [n]$ and each k-clause $x_{i_1} \wedge \bar{x}_{i_2} \wedge \ldots \wedge x_{i_k}$ is represented by a hyperedge $\{i_1, \ldots, i_k\} \subset V = [n]$. Furthermore each such hyperedge carries an additional label that prescribes which variables appear with a negation (there are thus 2^k possible labels).

We are now in the situation where one starts with a (labeled) hypergraph $H_0 = ([n], \emptyset)$, and as the intensity $c = M/n$ gets larger, more and more labeled hyperedges appear randomly, thus forming a random labeled hypergraph H_c. The event we consider is $A = A(H) := \{$all assignments $\omega \in \{-1, 1\}^n$ realize the k-DNF function associated to $H\}$. We clearly have that $c \mapsto \mathbb{P}[H_c \in A]$ is increasing in c. Put in this way, the present setting is similar to the celebrated sharp threshold theorem on the connectedness of Erdős–Renyi graphs $G(n, p)$ around $p_c \sim \frac{\log n}{n}$. In some sense Friedgut extends in (Fri99) the study of sharp threshold of graph properties to the case of (labeled) hypergraphs. Because, even with graphs, it is not correct that all graph properties have a sharp threshold (e.g., the property "containing a triangle" does not have a sharp threshold in the generalized sense that the size of the phase transition must be negligible w.r.t. to the value of critical point), a significant amount of work in (Fri99) is devoted to finding necessary conditions for "graph-like" properties of hypergraphs to exhibit a sharp threshold. The k-SAT problem is shown to fall into this class. See (Fri99) and (O'D14) for much more on this topic.

12.6 Noise sensitivity with respect to other "noises"

In the *traditional* noise sensitivity studied in this book, the noise was of a specific type, namely, each bit is re-randomized independently. One can easily imagine other types of noises that one could use. In this section, we discuss three such variants.

12.6.1 (Fixed-size)-noise sensitivity

In (BKS99), the following noise was also introduced. One has an integer parameter $q \in [0, n]$ and given $\omega \in \Omega_n$, one chooses q of the n bits uniformly at random and flips these bits. We call the resulting configuration ω_q.

Definition 12.20 The sequence $\{f_n\}$ defined on Ω_n is **(fixed-size)-noise sensitive** if for every $\epsilon > 0$ and any sequence $q_n \in (\epsilon n, (1 - \epsilon)n)$

$$\lim_{n \to \infty} \mathbb{E}[f_n(\omega) f_n(\omega_{q_n})] - \mathbb{E}[f_n(\omega)]^2 = 0. \tag{12.7}$$

Observe that Parity is a simple example that is noise sensitive but clearly not (fixed-size)-noise sensitive. The following result is proved in (BKS99).

Proposition 12.21 (BKS99)   *Consider a sequence of Boolean functions* f_n.
(i) $\{f_n\}$ *is (fixed set)-noise sensitive if and only if, for any* $k \geq 1$,

$$\sum_{m \in \{1,2,\dots,k\} \cup \{n-k,n-k+1,\dots,n\}} \sum_{|S|=m} \hat{f}_n(S)^2 \xrightarrow[n \to \infty]{} 0.$$

(ii) $\lim_n \sum_k \mathbf{I}_k(f_n)^2 = 0$ *implies that* $\{f_n\}$ *is (fixed set)-noise sensitive.*

The following corollary is immediate.

Corollary 12.22   *1. (Fixed-size)-noise sensitivity implies noise sensitivity. 2. For monotone functions, noise sensitivity and (fixed-size)-noise sensitivity are equivalent.*

12.6.2 Exclusion sensitivity

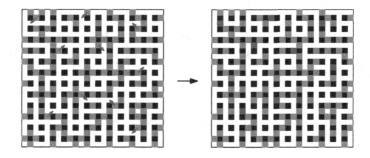

Figure 12.1 Illustration of a perturbation under the exclusion process.

In this model, ω is chosen uniformly at random as usual but now the *number* of 1's stays constant but the 1's move around under an exclusion process where the bits are considered to be the vertices of a graph. See Figure 12.1. A fairly thorough study of this model is carried out in (BGS13) to which we direct interested readers. We do not give a precise definitions here but do mention that as a consequence of the results in (BGS13), we have that exclusion sensitivity for the complete graph implies (fixed-size)-noise sensitivity. On the other hand, the Boolean function, which is 1 if and only if $\lfloor N(\omega)/\log n \rfloor$ is even (where $N(\omega)$ is the number of 1's), is easily seen to be (fixed-size)-noise sensitive but not exclusion sensitive. It is also shown in this paper that for monotone functions exclusion sensitivity for the complete graph is equivalent to noise sensitivity. For our main example of percolation crossings, exclusion sensitivity with respect to various intermediate range exclusion processes is also proved.

12.6.3 Percolation crossings in the Boolean model

In (ABGM13), one studies noise sensitivity of crossing events in the critical Poisson Boolean model. In this model, one has a Poisson process in the plane with intensity λ and one places balls of radius one around each of the Poisson points. One then looks if the collection of balls has an infinite component. The authors take λ to be the critical intensity λ_c and consider the event of a left–right crossing in a large box. The *noise* that is added to the system is as follows. Given a small number ϵ, one removes each point of the original Poisson process independently with probability ϵ and then adds to the system an independent Poisson realization with intensity $\epsilon \lambda_c$. The final configuration is of course also a Poisson realization with intensity λ_c. The authors show that the system is noise sensitive in that for each fixed $\epsilon > 0$, the events that the original realization has a left–right crossing and that the perturbed realization has a left–right crossing are asymptotically independent as the box size goes to ∞.

12.6.4 Noise sensitivity under partial noise

Another noise variant that was suggested in (BKS99) is where a specified subset of the bits is re-randomized with probability ϵ independently while the remaining bits are unchanged. This is discussed in (GPS10). An interesting example is when we consider percolation crossings and re-randomize only the vertical edges with probability 1. It is proved in

(GPS10) that noise sensitivity still results answering a question in (BKS99). Some general results concerning this type of noise are obtained in (GPS10).

12.7 Juntas and noise resistance

The following definition is central in this section.

Definition 12.23 A Boolean function f on Ω_n is a (ϵ, k)**-junta** if there exists a Boolean function g on Ω_n which depends on at most k variables such that $\mathbb{P}(f \neq g) \leq \epsilon$.

Juntas are viewed as *noise resistant* because it is immediate to check that if f is a (ϵ, k)**-junta**, then $\mathbb{P}[f(\omega) \neq f(\omega_\delta)] \leq 2\epsilon + k\delta/2$.

It is trivial that if f is a Boolean function that depends only on k bits (i.e., is a $(0, k)$-junta), then $\mathbf{I}(f)$ is at most k. The converse is false because there is a sequence of Boolean functions, whose variances are bounded away from 0, which depend on all the variables (which is equivalent to saying that each variable as positive influence) but such that the total influence is at most 2. We should therefore ask instead if a function with *small* total influence is a junta. The following theorem of Friedgut, which we will not prove, tells us that this is the case.

Theorem 12.24 (Fri98) *Let f be a Boolean function of n variables with $\mathbf{I}(f) \leq k$. Then for any $\epsilon > 0$, we have that f is a $(\epsilon, e^{\frac{3k}{\epsilon}})$-junta.*

The crucial point is of course that the size of the junta, $e^{\frac{3k}{\epsilon}}$, does not depend on n but only on the total influence and the desired degree of approximation of the junta. It is also shown in (Fri98) that the above theorem is sharp up to the constant 3 in the exponent, meaning that if 3 is replaced by a small constant, the result becomes false.

It turns out that Bourgain obtained a far-reaching extension of this which we will also not prove.

Theorem 12.25 (Bou02) *For all $\eta, \epsilon > 0$, there exists a $c_{\eta, \epsilon}$ such that if f is a Boolean function of n variables mapping into $\{\pm 1\}$ satisfying for some $k \geq 1$*

$$\sum_{|S| \geq k} \hat{f}(S)^2 \leq \frac{c_{\eta, \epsilon}}{k^{\frac{1}{2} + \eta}},$$

then f is a $(\epsilon, k10^k)$-junta.

Corollary 12.26 *For all $C, \delta, \epsilon > 0$, there exists $L = L(C, \delta, \epsilon)$ so that if f*

is a Boolean function of n variables mapping into $\{\pm 1\}$ *satisfying for all k*

$$\sum_{|S| \geq k} \hat{f}(S)^2 \leq \frac{C}{k^{\frac{1}{2}+\delta}},$$

then f is a (ϵ, L)-*junta. In particular, by Markov's inequality, f is a* (ϵ, L)-*junta provided that the* $\frac{1}{2} + \delta$-*moment of the spectral sample,* $\hat{Q}(|\mathscr{S}|^{\frac{1}{2}+\delta})$, *is at most C.*

Proof Let $\eta := \delta/2$. Choose k so that $C/k^{\eta} \leq c_{\eta,\epsilon}$. For this specific value of k, we have

$$\sum_{|S| \geq k} \hat{f}(S)^2 \leq \frac{c_{\eta,\epsilon}}{k^{\frac{1}{2}+\eta}}$$

and hence by Theorem 12.25 f is a $(\epsilon, k10^k)$-junta. Since k depends only on C, δ and ϵ, the theorem is proved. $\qquad \square$

Remark (1) By taking δ to be $1/2$ in the last statement of Corollary 12.26, you immediately obtain Theorem 12.24 without the specified dependence of the size of the junta on the total influence and ϵ but of course with the independence on n.

(2) Corollary 12.26 can be seen to be essentially sharp in the sense that the positive δ cannot be removed. To see this, one can consider the majority functions M_n. It is clear that these are "not juntas" while at the same time, it is known that for some constant C, one has that for each k and n

$$\sum_{|S| \geq k} \hat{M}_n(S)^2 \leq \frac{C}{k^{\frac{1}{2}}}.$$

(3) An exercise shows that $\sum_{|S_f| \geq 1/\epsilon} \hat{f}(S)^2 \leq O(1)\mathbb{P}[f(\omega) \neq f(\omega_\epsilon)]$. It follows that $\mathbb{P}[f(\omega) \neq f(\omega_\epsilon)] \leq \epsilon^{1/2+\delta}$ for a fixed $\delta > 0$ and all small $\epsilon > 0$ implies, by Corollary 12.26, that f is an appropriate junta. Note again that the majority functions are such that $\mathbb{P}[f(\omega) \neq f(\omega_\epsilon)]$ is of the order of $\epsilon^{1/2}$. Finally, observe that the assumption that the total influence is at most k immediately yields $\mathbb{P}[f(\omega) \neq f(\omega_\epsilon)] \leq k\epsilon$, a much stronger assumption than that this is at most $\epsilon^{1/2+\delta}$.

(4) There is no reasonable converse of Corollary 12.26 at all because there exists a sequence $\{f_n\}$ of Boolean functions such that $\hat{Q}(|\mathscr{S}_{f_n}|^\delta)$ approaches infinity for each positive δ but such that for each $\epsilon > 0$, for all large n, f_n is a $(\epsilon, 1)$-junta. For example, as readers can check, one can let f_n be defined on $1 + n + 2^{2^n}$ bits, which is x_1 unless $x_2, \ldots, x_n$ are all 1, in which case it is the mod 2 sum of all the bits except $x_2, \ldots, x_n$.

12.8 Social choice and Arrow's impossibility theorem

In this section, we are interested in *voting schemes* where instead of two candidates, one has three candidates in an election denoted by A, B, and C and one has n voters. We will assume that each voter $i \in [n]$ returns three bits $x_i, y_i, z_i \in \{-1, 1\}$ that encode his/her own preference respectively, among candidates A/B, B/C, and C/A. This way, if voter i returns, for example $(1, 1, -1)$, this means that voter i has the following ranking in mind: $A > B > C$. In that case, if the outcome of the election would depend only on voter i (i^{th} Dictator), candidate A would get elected. We will assume in what follows that all voters are *rational* voters in the sense that they will not return the rankings $(1, 1, 1)$ or $(-1, -1, -1)$ which would lead to the situations $A > B > C > A$ or $A < B < C < A$. Therefore, the possible votes are in the set $K_3 := \{-1, 1\}^3 \setminus \{(1, 1, 1), (-1, -1, -1)\}$, which is of size $2^3 - 2 = 6$. Furthermore, the result of a voting with n voters will be a certain configuration $\omega_n = (x_i, y_i, z_i)_{i \in [n]} \in K_3^n \subset \Omega_3^n$. At this point, one still has to choose a *voting scheme*, that is, a way to decide which candidate is elected. In this setting, one natural way to proceed as suggested long ago by Condorcet is to fix a certain Boolean function $f : \{-1, 1\}^n \to \{-1, 1\}$ and to consider the triple

$$R = (f(x_1, \ldots, x_n), f(y_1, \ldots, y_n), f(z_1, \ldots, z_n)) \in \{-1, 1\}^3$$

This voting scheme is reasonable if one can make sure that this triple will always lie in the set of admissible votes K_3. Unfortunately, it turns out that unless f depends only on one voter (i.e., is a Dictator or an anti-Dictator), it can be shown that even though all voters vote in K_3, there will exist configurations that will lead to an *irrational outcome* in K_3^c. This is the content of the following celebrated Arrow's Impossibility Theorem, where we denote $(x_1, \ldots, x_n)$ by x and similarly for y and z.

Theorem 12.27 (Arrow's impossibility theorem) *Let $n \geq 2$ and suppose $f : \Omega_n \to \{-1, 1\}$ is any Boolean function that does not depend on a single voter $i \in [n]$. Then there exists a configuration $\omega_n \in K_3^n$ so that $R = (f(x), f(y), f(z)) \notin K_3$.*

The reason why this theorem, which solves Condorcet's paradox appears in this book, is the fact that there exists a beautiful proof due to Gil Kalai (Kal02) that relies on Fourier analysis.

Proof (Gil Kalai)

We want to check whether the triple $(f(x), f(y), f(z))$ always lies in the set of rational outcomes K_3 or not. Note that if (X, Y, Z) is any point in $\{-1, 1\}^3 \supset K_3$, then the Boolean function

$$g(X, Y, Z) := \frac{3}{4} - \frac{1}{4}XY - \frac{1}{4}XZ - \frac{1}{4}YZ,$$

returns 1 if $(X, Y, Z) \in K_3$ while it returns 0 otherwise. Therefore if $\mathbb{P}$ denotes here the uniform measure on configurations $(x_i, y_i, z_i)_{i \in [n]} \in K_3^n$, one has for any Boolean function $f: \{-1, 1\}^n \to \{-1, 1\}$

$$\mathbb{P}[(f(x), f(y), f(z)) \in K_3] = \frac{3}{4} - \frac{3}{4}\mathbb{E}[f(x)f(y)],$$

where we used the symmetry among the variables x, y, and z.

Notice now that if the configuration $(x_i, y_i, z_i)_{i \in [n]}$ is chosen uniformly in K_3^n, then the projected vectors $(x_i)_i, (y_i)_i, (z_i)_i$ are each uniform random configurations in the hypercube Ω_n. Yet, if one is interested in the coupling of two of them, say $(x_i, y_i)_{i \in [n]}$, it is not correct that they form independent vectors in $\{-1, 1\}^n$. Their correlation is given by $\mathbb{E}[x_1 y_1]$, which is simply seen to be equal to $-1/3$. In other words, among rational outcomes in K_3, the variables are negatively correlated. We have now enough information to carry out the above computation:

$$\begin{aligned}
\mathbb{P}[(f(x), f(y), f(z)) \in K_3] &= \frac{3}{4} - \frac{3}{4}\mathbb{E}[f(x)f(y)] \\
&= \frac{3}{4} - \frac{3}{4}\sum_{S,S'} \hat{f}(S)\hat{f}(S')\mathbb{E}[\prod_{i \in S, j \in S'} x_i y_j] \\
&= \frac{3}{4} - \frac{3}{4}\sum_{S} \hat{f}(S)^2(-\frac{1}{3})^{|S|}.
\end{aligned}$$

In order that the voting always produces a rational outcome, clearly f cannot be a constant because then f would be identically equal to 1 or -1 and this would always lead to an irrational outcome. However, in fact, one has the stronger statement that $\hat{f}(\emptyset)$ must be equal to 0 if the voting always produces a rational outcome. To see this, first note that if $p :=$

$\mathbb{P}[(f(x), f(y), f(z)) \in K_3]$, one would have

$$p = 3/4 - 3/4\hat{f}(\emptyset)^2 - 3/4 \sum_{|S| \geq 1} \hat{f}(S)^2(-\frac{1}{3})^{|S|}$$

$$\leq 3/4 - 3/4\hat{f}(\emptyset)^2 + 1/4 \sum_{|S| \geq 1} \hat{f}(S)^2$$

$$= 3/4 - 3/4\hat{f}(\emptyset)^2 + 1/4(1 - \hat{f}(\emptyset)^2) \text{ (by Parseval)}$$

$$= 1 - \frac{\hat{f}(\emptyset)^2}{2}.$$

Therefore, if one is looking at a Boolean functions f such that $p = 1$, we must have $\hat{f}(\emptyset) = 0$, as claimed.

Let us now characterize the Boolean functions f with $\hat{f}(\emptyset) = 0$ such that

$$p = 1 = \frac{3}{4} - \frac{3}{4} \sum_{|S| \geq 1} \hat{f}(S)^2(-\frac{1}{3})^{|S|}.$$

Using once again the fact that $f \in \{-1, 1\}$ and thus $\sum_S \hat{f}(S)^2 = 1$ by Parseval, it is immediate to see that all the nonzero Fourier coefficients of f must correspond to singletons. Hence $f = \sum_k a_k x_{i_k}$ with $\sum a_k^2 = 1$. It is an easy exercise to check that the only such functions that are Boolean (i.e., with values in $\{-1, 1\}$) are Dictators $f = x_{i_1}$ and anti-Dictators $f = -x_{i_1}$ for some $i_1 \in [n]$, which thus ends the proof of Theorem 12.27. □

Remark The advantage of this Fourier proof by Kalai (besides the fact that it is a beautiful argument) lies in the fact that it gives sharp quantitative bounds if one relaxes slightly the problem and asks instead how a Boolean function $f: \Omega_n \to \{-1, 1\}$ must be chosen if one wants $p = \mathbb{P}[(f(x), f(y), f(z)) \in K_3]$ to be not exactly 1 but very close to it, say $1 - \delta$ for some small parameter δ. It is possible to show that any Boolean function f satisfying the above property is such that there is always a Dictator (or an anti-Dictator) $g: \Omega_n \to \{-1, 1\}$ that approximates f very well in the sense that $\mathbb{P}[f \neq g] \leq O(\delta)$. See (Kal02) for more details.

12.9 Non-interactive correlation distillation and Borell's reverse hypercontractivity theorem

Letting T_ρ be as in Chapter 5, an exercise shows that for any function f on the hypercube mapping into $\{0, 1\}$, we have that

$$\mathbb{P}[f(\omega) = f(\omega_\epsilon)] = \mathbb{E}[(T_{1-(1-\sqrt{1-\epsilon})}f)^2] + \mathbb{E}[(T_{1-(1-\sqrt{1-\epsilon})}(1 - f))^2].$$

It turns out to be interesting to consider higher moments of the above functions and in particular it turns out that the expression

$$\mathbb{E}[(T_{1-\epsilon}f)^k] + \mathbb{E}[(T_{1-\epsilon}(1-f))^k]$$

has an interesting probabilistic interpretation related to a concept called *non-interactive correlation distillation*. Imagine a source of n i.i.d. coin flips that are sent to k different people. Given the realization of the source, each of the nk bits that are transmitted is re-randomized independently with probability ϵ. The k people want to "flip a fair coin" based on what they receive in such a way that the probability that all k players get the same outcome is maximized. This question is motivated by cryptography considerations. More precisely, let f be a function from sequences of length n into $\{0, 1\}$, which takes the value 1 half the time. So the output of the kth player is f applied to the sequence that the kth player received. The term non-interactive correlation distillation comes from the fact that the k players want to jointly distill out the correlation between their bit string and the original bit string but that no interaction is allowed between the k players. We assume that all players use the same function f. The following proposition is easy to prove and left to the reader.

Proposition 12.28 *The probability that the coins that the k players "toss" by applying f to their received bit string all have the same outcome is*

$$\mathbb{E}[(T_{1-\epsilon}f)^k] + \mathbb{E}[(T_{1-\epsilon}(1-f))^k].$$

It was shown in (MO05) that when k is 2 or 3, then the function f that maximizes the above probability is a Dictator function. The Fourier argument which shows that the (± 1) Dictator function is the most stable mean 0 function shows that the $(0, 1)$ Dictator function is the unique optimal function for $k = 2$ and also that it uniquely maximizes the expected number of pairs of players that have the same outcome. One can see that for $k = 3$, this is equivalent to maximizing the probability that all the players have the same outcome. However, as shown in this paper, Dictator is not always the optimal function. They show that for n and ϵ fixed, for all k sufficiently large, the optimal f is to use the Majority function. Many other results are obtained in (MO05).

Let $p(n, k, \epsilon)$ be the probability that all players have the same outcome under the optimal function f under the parameters n, k and ϵ. Note that $\lim_n p(n, k, \epsilon)$ trivially exists, which we denote by $p(\infty, k, \epsilon)$. In a follow-up paper on this model, (MOR$^+$06), the following theorem was shown. Although the theorem might seem *obvious*, it seems nontrivial to prove.

Theorem 12.29 *For all $\epsilon > 0$, $\lim_k p(\infty, k, \epsilon) = 0$.*

One of the key techniques needed to prove the above result is a reverse hypercontractivity theorem due to C. Borell contained in (Bor82) and whose statement is the following.

Theorem 12.30 *Let $f: \{\pm 1\}^n \to \mathbb{R}^{\geq 0}$ be a nonnegative function and let $-\infty < q \leq p \leq 1$. Then*

$$\|T_\rho f\|_q \geq \|f\|_p \qquad \text{for all } 0 \leq \rho \leq (1-p)^{1/2}/(1-q)^{1/2}. \tag{12.8}$$

The proof of Theorem 12.30 can be found in either (Bor82) or in (MOR$^+$06). Theorem 12.29 was the first application of the reverse hypercontractivity theorem to theoretical computer science. Since then, other applications have been found.

12.10 Deterministic and randomized complexity for Boolean functions

In this section, we discuss the important concepts of deterministic and randomized complexity for general Boolean functions. The main result of this section is proved in (OSSS05); this result proves a general lower bound on the randomized complexity for nontrivial transitive monotone Boolean functions. Such a lower bound was previously proved for the special case of monotone graph properties. A key step in the proof is to obtain an inequality that relates influence with the complexity of the function. Then, we will see a major open question concerning randomized complexity for nontrivial monotone graph properties. Finally, at the end of this section, we also give another application to arms events in critical percolation in the same spirit as the inequality obtained in Section 8.5 in Chapter 8.

Let us now formally introduce the relevant concepts. For background concerning these topics, in addition to (OSSS05), see also (Haj92). We consider a Boolean function f. As in Section 8.2 of Chapter 8, an algorithm deterministically queries the bits, where the bit that is queried may depend on the values of the previous bits seen. It is assumed that any algorithm stops as soon as the output of the function f can be determined from the given information. In theoretical computer science, this is referred to as a *decision tree* but we won't use this terminology here. Given an algorithm A for f and $\omega \in \{-1, 1\}^n$, we let $c(A, \omega)$ be the number of queries that the algorithm A performs when the input is ω. Let $c(A) := \max_\omega c(A, \omega)$ be the maximum number of queries that A makes which we view as the cost of A. Finally, we have the following important definition.

Definition 12.31   The **deterministic complexity** of a Boolean function f, denoted by $D(f)$, is the minimum of $c(A)$ over all algorithms A for f.

Remark It is not so hard to show that for a Boolean function f of n variables, $D(f) = n$ if and only there is an adversary who can give answers to bit queries that always forces all the bits to be queried; this latter property is called *evasiveness*.

Although we are primarily interested in the concept of randomized complexity, to be given soon, we do mention one theorem concerning deterministic complexity, which, to our taste, is quite interesting. It is proved in (RV75).

Theorem 12.32   *Consider a nontrivial transitive monotone Boolean function f of n variables. If n is a prime power, then $D(f) = n$.*

This theorem was derived to then prove (in the same paper) the Aanderaa–Rosenberg Conjecture which states that the deterministic complexity for any nontrivial monotone graph property on n vertices is $\Omega(n^2)$.

A randomized algorithm is a variant of an algorithm where the next bit chosen may depend not only on the bit values seen so far but also on exterior randomness. This is equivalent to choosing a (deterministic) algorithm at random according to some distribution. We denote a randomized algorithm by $\tilde{A}$. Given a randomized algorithm $\tilde{A}$ for f and $\omega \in \{-1, 1\}^n$, we let $c(\tilde{A}, \omega)$ be the *expected* number of queries that the randomized algorithm $\tilde{A}$ performs on the input ω. Let $c(\tilde{A}) := \max_\omega c(\tilde{A}, \omega)$ be the maximum (over the input strings) number of queries that $\tilde{A}$ makes on average, which we view as the cost of $\tilde{A}$. Finally, we have the following important definition.

Definition 12.33   The **randomized complexity** of a Boolean function f, denoted by $R(f)$, is the minimum of $c(\tilde{A})$ over all randomized algorithms $\tilde{A}$ for f.

To see the difference between deterministic complexity and randomized complexity, consider majority on 3 bits. It is easy to see that $D(f) = 3$ but $R(f) < 3$. These concepts are very related to those in Section 8.4 in Chapter 8. However, note that in that section, the input bits are assumed to be i.i.d. uniform while here we are considering worst case input, which is an essential difference.

Although perhaps less central here, we introduce a third complexity measure that is defined in terms of a concept we already saw, namely that of a witness.

Definition 12.34   The **nondeterministic complexity** of a Boolean function f, denoted by $N(f)$, is $\max_{\omega} w(\omega)$, where $w(\omega)$ is given in Definition 8.11.

Because at the end of any algorithm, the queried bits must be a witness, we trivially have $D(f) \geq R(f) \geq N(f)$. Clearly $N(f)$ could be n such as in the case of Parity. Although we do not prove it here, the following result is due to M. Blum and others.

$$D(f) \leq (N(f))^2. \tag{12.9}$$

Interestingly, this is sharp. If one takes a binary tree of height $2k$ with the leaves corresponding to the input bits and lets the even levels correspond to an AND-function while the odd levels correspond to an OR-function, then it is not so hard to show that the resulting Boolean function f satisfies $D(f) = 2^{2k}$ while $N(f) = 2^k$. Another such example is if the input is a 0–1 square matrix and the output is a 1 if there is row consisting of all 1's.

We now stick only to the first two notions of complexity. Note that an immediate consequence of (12.9) is that $D(f) \leq (R(f))^2$. For general monotone transitive Boolean functions, $R(f)$ can be much smaller than $D(f)$. For example, for Iterated 3-Majority iterated k times, while Theorem 12.32 tells us that $D(f)$ is 3^k (this is also easy to see directly), it is easy to show that $R(f)$ is at most $(8/3)^k$. (What the smallest number that can replace $8/3$ here for all large k is is an open question.)

The following question seems to be one of the major questions in the field. It concerns a variant of the Aanderaa–Rosenberg Conjecture for randomized complexity.

Conjecture 12.2   *(Karp) For Boolean functions f corresponding to a nontrivial monotone graph property on n vertices, we have that $R(f) = \Omega(n^2)$.*

There have been a number of results proving weaker versions of this; here we simply mention the result in (Haj91), which states that in this graph property context one has $R(f) \geq \Omega(n^{4/3})$. This is the best result in terms of the power of n.

We now move to one of the two main theorems of this section. We include the proof here because it uses many of the concepts already introduced in the book (and because we find it very interesting).

Theorem 12.35 ((OSSS05))   *If f is a monotone, transitive nontrivial Boolean function, then*

$$R(f) \geq n^{2/3}.$$

This result is much more general than that in (Haj91) mentioned previously and the proof technique completely different. In this regard, it is quite amazing that the exact same power of n is obtained.

One of the key steps in proving this result is the following interesting result, which can be viewed as a strengthening of Poincaré inequality (Theorem 1.13).

Theorem 12.36 (OSSS05) *Let f be a Boolean function mapping into $\{\pm 1\}$, A a randomized algorithm for f and $p \in (0, 1)$. Then*

$$\mathrm{Var}_{\pi_p}(f) \le 4p(1-p) \sum_i \delta_i^p \mathbf{I}_i^p(f)$$

where δ_i^p is the probability that i is queried by the algorithm A when the input distribution is π_p and $\mathbf{I}_i^p(f)$ is the level p influence of bit i.

Proof It suffices by linearity to prove the result when A is a deterministic algorithm. In this proof, we will change our previous notation slightly. Given ω and i, we let ω^i be ω except re-randomized in position i (as opposed to flipping the ith coordinate as we did earlier). It is then trivial to check that $\mathbb{P}(f(\omega) \ne f(\omega^i)) = 2p(1-p)\mathbf{I}_i^p(f)$ Next, it is elementary to check that $\mathrm{Var}_{\pi_p}(f) = 2\mathbb{P}(f(x) \ne f(y))$, where x and y are chosen independently from π_p. It follows that the inequality to be proved is

$$\mathbb{P}(f(x) \ne f(y)) \le \sum_i \delta_i^p \mathbb{P}(f(\omega) \ne f(\omega^i)),$$

where x and y are chosen independently from π_p. We now do this.

Applying the algorithm A to x, we let $i_1, \dots, i_s$ be the bits queried by A in the order they are probed. For $t = 0, \dots s-1$, let u_t be x on $i_{t+1}, \dots, i_s$ and y on the rest. For $t \ge s$, let u_t be y. Expressed in words, u_t is obtained by taking the configuration x and replacing all the values of the bits that have been probed up to step t *or* are never probed by their corresponding y values.

Observe that $f(x) = f(u_0)$ and therefore that

$$\{f(x) \ne f(y)\} \subseteq \cup_{t=1}^n \{f(u_{t-1}) \ne f(u_t)\} = \cup_{t=1}^n \cup_{i=1}^n \{f(u_{t-1}) \ne f(u_t), i_t = i\}$$

where i_t is (arbitrarily) defined to be 0 for $t > s$.

For each t and i, by first conditioning on all the information obtained after $t - 1$ steps of A, it is not hard to see that $\mathbb{P}(\{f(u_{t-1}) \ne f(u_t), i_t = i\}) = \mathbb{P}(i_t = i)\mathbb{P}(f(\omega) \ne f(\omega^i))$. By summing up first over t, noting that $\sum_t \mathbb{P}(i_t = i) = \delta_i^p$ and then summing over i, we obtain the result. $\square$

Remark By linearity, Theorem 12.36 holds for randomized algorithms as well.

Before stating the next lemma, we need another definition.

Definition 12.37 The **level p complexity** of a Boolean function f, denoted by $D_p(f)$, is the minimum over all (deterministic) algorithms A for f of the expected number of questions that are asked when the input has distribution π_p.

Observe that this definition is unchanged if we allow randomized algorithms and that $D_{1/2}$ is the notion that we were dealing with in Section 8.4 of Chapter 8. The following reduces to Theorem 8.8 when $p = 1/2$; because the proof is more or less the same, we skip it.

Lemma 12.38 (OS07) *For monotone Boolean functions f, $D_p(f) \geq 4p(1 - p)(\sum_i \mathbf{I}_i^p(f))^2$.*

Proof of Theorem 12.35. Fix a monotone nontrivial transitive Boolean function. Fix p arbitrarily and choose an algorithm A minimizing the expected number of questions that are asked when the input has distribution π_p. Theorem 12.36 and transitivity yield $\mathrm{Var}_{\pi_p}(f) \leq 4p(1 - p)D_p(\sum_i \mathbf{I}_i^p(f))/n$. Lemma 12.38 tells us that the right-hand side is at most $2\sqrt{(1 - p)p}D_p^{3/2}/n$. This gives

$$D_p \geq n^{2/3}(4p(1 - p))^{-1/3}\mathrm{Var}_{\pi_p}(f)^{2/3}.$$

Since f is monotone and nontrivial, we can choose p so that $\mathrm{Var}_{\pi_p}(f) = 1$. Since $R(f) \geq D_p$ for all p, the result is proved. $\qquad\square$

We lastly give an interesting corollary of Theorem 12.36 that immediately yields Theorem 8.9.

Corollary 12.39 *For any Boolean function f and any $p \in (0, 1)$, we have*

$$\max_i \mathbf{I}_i^p(f) \geq \mathrm{Var}_{\pi_p}(f)/(D_p 4p(1 - p))(\geq \mathrm{Var}_{\pi_p}(f)/(R(f)4p(1 - p))).$$

This yields a lower bound on the maximum influence in terms of the algorithmic complexity of the function; previous results of this type were much weaker.

Let us end this section with another surprising application to arms exponent in critical percolation (in the same spirit as in Section 8.5 of Chapter 8). Indeed we shall see that Theorem 12.36 implies the following nontrivial relation:

Theorem 12.40 *There is a constant $C > 0$ such that for critical percolation on $\mathbb{Z}^2$,*

$$\alpha_2(R)\alpha_4(R) \geq C\alpha_5(R).$$

Remark (1) Note that the corresponding inequality is of course known for critical percolation on the triangular lattice because it would correspond to $R^{-1/4+o(1)}R^{-5/4+o(1)} \geq CR^{-2}$, but it is far from being obvious using standard arguments on the square grid $\mathbb{Z}^2$. In fact, we believe that such a relation between arms exponents was not known earlier.

(2) Note also that the weaker inequality $\alpha_1(R)\alpha_4(R) \geq \alpha_5(R)$ follows easily from Reimer's inequality (see equation (6.7) or (Re00)).

Sketch of proof

Let us apply Theorem 12.36 with $p = p_c(\mathbb{Z}^2) = 1/2$ to the left–right crossing event f_R in a $R \times R$ square with the same randomized algorithm A as in Section 8.3 of Chapter 8.

Using similar techniques as in Chapter 6 (i.e., dealing with boundary issues), it can be shown that there is a positive constant $B < \infty$ such that

$$\sum_{x\in[0,R]^2} \delta_x \mathbf{I}_x(f_R) \leq B \sum_{x\in[R/4,3R/4]^2} \delta_x \mathbf{I}_x(f_R). \tag{12.10}$$

Note that the boundary issues are similar but somewhat different here: the important observation is to notice that both δ_x and $\mathbf{I}_x(f_R)$ get smaller near the boundary (and so does their product). It is also clear that for any $x \in [R/4, 3R/4]^2$, $\delta_x \leq \alpha_2(R/4)$ and $\mathbf{I}_x(f_R) \leq \alpha_4(R/4)$. This, (12.10), and Theorem 12.36 imply that

$$\Omega(1) \leq \text{Var}(f_R) \leq \sum_x \delta_x \mathbf{I}_x(f_R)$$
$$\leq B \sum_{x\in[R/4,3R/4]^2} \delta_x \mathbf{I}_x(f_R)$$
$$\leq B(R/2)^2 \alpha_2(R/4)\alpha_4(R/4)$$
$$\leq O(1)R^2 \alpha_2(R)\alpha_4(R),$$

where we used quasi-multiplicativity (Proposition 2.3). We conclude the proof using Theorem 6.4 on the five-arms exponent in $\mathbb{Z}^2$.

$\square$

Some exercises related to Section 12.10

12.1 Determine $R(f)$ for Majority on 3 bits.

12.2 Letting $\max\{k : \exists S \text{ such that } |S| = k, \hat{f}(S) \neq 0\}$ be the degree of a Boolean function f, show that $D(f)$ is at least the degree of f.

12.3 Show that if f is reasonably balanced on n bits, then the revealment is at least of order $1/n^{1/2}$. Give a version of this for witnesses.

12.4 Show that if f is reasonably balanced on n bits and is monotone, then the revealment is at least of order $1/n^{1/3}$.

12.11 Erdös–Rényi random graphs and strong noise sensitivity

Throughout most of this book, the parameter p has mostly been taken to be $1/2$. In many important cases, it is natural to let p_n vary with n such as in the Erdős–Rényi random graph model $\mathcal{G}(n,p)$. Note that noise sensitivity, when one assumes the functions are nondegenerate, is equivalent to

$$\lim_{n\to\infty} \mathbb{P}[f_n(\omega_\epsilon) = 1 | f_n(\omega) = 1] - \mathbb{P}[f_n(\omega) = 1] = 0 \qquad (12.11)$$

where p_n is implicit in the expectation. It turns out that Theorem 1.21 is no longer true in this context. For example, if one considers the Erdős–Rényi random graph $\mathcal{G}(n, n^{-2/3})$ and f is the event of containing a K_4, then it is easy to see that (1.2) holds while noise sensitivity fails. (It turn out that the correct analog of (1.2) in the small p context is that the left–hand size of (1.2) when multiplied by p_n should go to 0; this then holds even for $\mathcal{G}(n, 1/n)$ with f being the event of containing a triangle with noise sensitivity again failing.) For varying p, Keller and Kindler (KK13) have a result that extends Theorem 1.21 into the regime $p_n = 1/n^{o(1)}$. However, in the interesting regimes for the Erdős–Rényi random graph model $\mathcal{G}(n,p)$, for example, when connectivity, containing a Hamiltonian cycle or various events involving the critical giant component are nondegenerate, p_n is much smaller than $1/n^{o(1)}$ and so this is not covered by the result in (KK13).

In (LS), a strengthening of the notion of noise sensitivity for monotone functions is studied and from this, various properties for $\mathcal{G}(n,p)$ are proved to be noise sensitive. We assume monotonicity for the rest of this section and for simplicity, we assume, only in this section, that our Boolean functions map $\{0,1\}^n$ into $\{0,1\}$. Recall that a **1-witness** is a minimal subset of the variables with the property that if all the bits in this subset are 1, then the function is guaranteed to be 1. Let $\mathcal{W}_1 = \mathcal{W}_1(f)$ denote the set of 1-witnesses of some monotone Boolean function f and similarly for $\mathcal{W}_0 = \mathcal{W}_0(f)$.

Definition 12.41 A sequence of monotone Boolean functions $f_n : \{0,1\}^n \to \{0,1\}$ is called **1-strongly noise sensitive** (STRSENS$_1$) w.r.t. (p_n) if for any

$\epsilon > 0$,

$$\lim_{n \to \infty} \max_{W \in \mathcal{W}_1} \mathbb{P}[f_n(\omega_\epsilon) = 1 \mid \omega_W \equiv 1] - \mathbb{P}[f_n(\omega) = 1] = 0. \tag{12.12}$$

0-strongly noise sensitive (STRSENS$_0$) is defined analogously. Equation (12.11) says that knowing $f_n(\omega) = 1$ gives us, for large n, almost no information about whether $f_n(\omega_\epsilon) = 1$. The event $\{f_n(\omega) = 1\}$ is the same as the event that $\omega_W \equiv 1$ for some $W \in \mathcal{W}_1$. If all the witnesses are similar, it might therefore seem at first that STRSENS$_1$ should be the same as noise sensitive but after further reflection, one sees that this is not the case.

It is easy to see that STRSENS$_1$ implies noise sensitive while the converse is not true. It is easy to show that Tribes is STRSENS$_1$; however, it is not STRSENS$_0$ showing that the above converse is false and that STRSENS$_1$ and STRSENS$_0$ are not equivalent. Interestingly, unlike noise sensitivity, it is possible that (12.12) holds for some ϵ and not for others. It can also happen (e.g., for Iterated 5-Majority but not for Iterated 3-Majority) that $\min_{W \in \mathcal{W}_1} \mathbb{P}[f_n(x^\epsilon) = 1 \mid x_W \equiv 1]$ approaches 1 as $n \to \infty$ for *fixed* ϵ, something that cannot occur with respect to the usual notion of noise sensitivity.

The size of the largest component for $\mathcal{G}(n, c/n)$ is of order n if $c > 1$, of order $n^{2/3}$ if $c = 1$, and of order $\log n$ if $c < 1$. It is known that there is a "critical window" $[1/n - c/n^{4/3}, 1/n + c/n^{4/3}]$ where this transition occurs. The size of the largest cycle when $p = 1/n$, when divided by $n^{1/3}$, has a nontrivial limiting distribution.

Theorem 12.42 (LS) *The event that there exists a cycle of length contained in $[n^{1/3}, 2n^{1/3}]$ is* STRSENS$_1$ *and hence is noise sensitive. In addition, one obtains (quantitative)* STRSENS$_1$ *when $\epsilon_n \gg n^{-1/3}$ and stability when $\epsilon_n \ll n^{-1/3}$.*

It is not hard to see that the "noise sensitivity exponent" here of $1/3$ matches exactly the critical window described above. Other events that one can prove are noise sensitivity are "connectivity" and "containing a Hamiltonian cycle," which are nondegenerate respectively at $\log n/n$ and $(\log n + \log \log n)/n$.

Theorem 12.43 (LS) *The events "connectivity" and "containing a Hamiltonian cycle" are each noise sensitive at the above values of p_n. In addition, one obtains (quantitative) noise sensitivity when $\epsilon_n \gg 1/\log n$ and stability when $\epsilon_n \ll 1/\log n$.*

The method here is to show that the events "minimum degree at least 1" and "minimum degree at least 2" are each STRSENS$_0$ (for similar reasons to why Tribes are STRSENS$_1$) and then to use the fact that these two events approximate very well the above two events.

We have seen in Problem 1.9 that for $p = 1/2$, clique containment is noise sensitive. In general, if H_n is a growing sequence of given graphs, the sequence of events "$H_n \subseteq \mathcal{G}(n, p)$" need not be noise sensitive as can be seen by letting H_n be $\log n$ disjoint edges.

A graph is called **strictly balanced** if its edge/vertex ratio is strictly larger than that of all of its subgraphs. The following provides some cases when noise sensitivity can be concluded for events of this form.

Theorem 12.44 (LS)   *(i) If H_n is strictly balanced with $1 \ll \ell_n \leq (\frac{\log n}{\log \log n})^{1/2}$ edges, then (f_n) is noise sensitive, and furthermore, it is* STRSENS$_1$.
(ii) There exists a sequence $\{H_n\}$ of strictly balanced graphs with $\ell_n \asymp \log n$ edges for which (f_n) is not noise sensitive.

12.12 Noise sensitivity and correlation with majority functions

The general theme of this section is the relationship between being noise sensitivity and being somewhat uncorrelated with all majority functions. The starting point of this section is the following proposition, which is not proved as it follows immediately from the stability of the Majority function.

Proposition 12.45   *Given a subset $K \subseteq [n]$, let M_K be the Boolean function on $\{0, 1\}^{[n]}$ which is just majority on the bits in K. (This function is 1 when there are more 1's than 0's in K, is -1 when there are more 0's than 1's in K and is 0 if a tie.) If $\{f_n\}$ is noise sensitive, then*

$$\lim_{n \to \infty} \sup_{K \subseteq [n]} E[f_n M_K] = 0.$$

One can ask for the converse of the above proposition. The answer is trivially no because if f_n is $\chi_{\{1,2\}}$ for each n, then f_n is uncorrelated with every M_K since the former is an even function and the latter is odd. The goal here is to show that for monotone functions, there is some type of converse. The first main result is the following.

Theorem 12.46 (BKS99)   *Let $\Lambda(f) := \max\{|\mathbb{E}(f M_K)| : K \subseteq [n]\}$. There exists C such that for all $f : \{0, 1\}^{[n]} \to \{0, 1\}$ which is monotone,*

$$\mathbf{H}(f) \leq C(\Lambda(f))^2 (1 - \log(\Lambda(f))) \log n.$$

Remarks (1). Since f is monotone, the FKG inequality tells us that

$\mathbb{E}(fM_K) \geq 0$.

(2). Theorem 12.46 states that if the maximum correlation of f_n with all majority functions goes to 0 slightly faster than $1/(\log n)^{1/2}$, then the sequence is noise sensitive provided the functions are monotone.

The first lemma needed in the proof of the above result is

Lemma 12.47 *There exist constants C_1, C_2 so that for all n and $\lambda \geq 0$, we have*

$$\frac{1}{2^n} \sum_{k \geq \frac{n+\lambda\sqrt{n}}{2}}^{n} \binom{n}{k}(2k-n) \leq C_1 \sqrt{n} e^{-C_2\lambda^2}.$$

Remark Note that if X is a binomial random variable with parameters n and $1/2$, then the left-hand side above is just

$$E[(2X-n)I_{\{X \geq \frac{n+\lambda\sqrt{n}}{2}\}}].$$

Proof We give only an outline. One can use the local central limit theorem to do this. However, we explain "why" it is true by easily showing the inequality when X is replaced by a normal random variable with the same mean and variance as X. (This suggests, because of the CLT theorem, that the result is true but does not prove it.)

Assuming X is exactly $Z\sqrt{n}/2 + n/2$ where Z is a standard normal, then $E[(2X-n)I_{\{X \geq \frac{n+\lambda\sqrt{n}}{2}\}}]$ becomes, after some easy algebra,

$$\sqrt{n}E[ZI_{\{Z \geq \lambda\}}].$$

The last expectation can be trivially computed exactly and it is $(1/\sqrt{2\pi})e^{-\lambda^2/2}$. □

We now need to define the influence of a variable for a function $f: \{0,1\}^{[n]} \to [0,1]$ which is monotone. We take $\mathbf{I}_i(f)$ to be defined to be

$$\mathbb{E}(f|x_i = 1) - \mathbb{E}(f|x_i = 0).$$

It is easy to check that if the image is $\{0,1\}$ and the function is monotone, then this agrees with our earlier definition. As before, the total influence, $\mathbf{I}(f)$, is defined to be $\sum_i \mathbf{I}_i(f)$.

Lemma 12.48 *There exists a constant C so that for all n and for all $f: \{0,1\}^{[n]} \to [0,1]$ which is monotone,*

$$\mathbf{I}(f) \leq C\sqrt{n}\mathbb{E}(fM_n)(1 + \sqrt{-\log(\mathbb{E}(fM_n))}).$$

Proof

Let $\overline{f}(k)$ be the average of f on the set $\{\sum x_i = k\}$; this is $\binom{n}{k}^{-1}\sum_{|x|=k} f(x)$. It is easy to see that

$$\mathbb{E}(fM_n) = 2^{-n}\sum_{k>n/2}\binom{n}{k}[\overline{f}(k) - \overline{f}(n-k)].$$

On the other hand,

$$\mathbf{I}(f) = 2^{-n}\sum_x\sum_j |f(x) - f(x^j)|$$

where x^j is x flipped at j. If f is now monotone, this is

$$2^{-n}2\sum_{(y,w):y\leq w,|w|=|y|+1}(f(w) - f(y)).$$

This is the same as

$$2^{-n+1}\sum_x f(x)|x| - f(x)(n-|x|)$$

since each x comes up as a w in the previous sum $|x|$ times and as a y, $n-|x|$ times. This simplifies to

$$2^{-n+1}\sum_x f(x)(2|x| - n)$$

$$= 2^{-n+1}\sum_{k=0}^{n}\binom{n}{k}\overline{f}(k)(2k-n)$$

$$= 2^{-n+1}\sum_{k>n/2}\binom{n}{k}[\overline{f}(k) - \overline{f}(n-k)](2k-n).$$

Given $\lambda > 0$, let $k(\lambda) = \frac{n+\lambda\sqrt{n}}{2}$. We have

$$\mathbf{I}(f)/2 = 2^{-n}\sum_{k>n/2}^{k(\lambda)}\binom{n}{k}[\overline{f}(k) - \overline{f}(n-k)](2k-n)$$

$$+ 2^{-n}\sum_{k>k(\lambda)}^{n}\binom{n}{k}[\overline{f}(k) - \overline{f}(n-k)](2k-n)$$

$$\leq \lambda\sqrt{n}\mathbb{E}[fM_n] + C_1\sqrt{n}e^{-C_2\lambda^2}$$

by Lemma 12.47. Setting

$$\lambda = (1/C_2^{1/2})\sqrt{-\log\mathbb{E}[fM_n]},$$

the claim is obtained. □

For $K \subseteq [n]$, we now let $\mathbf{I}_K(f) := \sum_{k \in K} \mathbf{I}_k(f)$.

Corollary 12.49 *There exists $C > 0$, such that if $f : \{0,1\}^{[n]} \to [0,1]$ is monotone, then for all $K \subseteq [n]$,*

$$\mathbf{I}_K(f) \le C \sqrt{|K|} E(fM_K)\left(1 + \sqrt{-\log(E(fM_K))}\right).$$

Proof Assume that $K = \{1, \ldots, m\}$. For $z \in \{0,1\}^m$, let

$$f_K(z) = \mathbb{E}[f | \omega = z \text{ on } K] = (1/2^{n-m}) \sum_{y \in \{0,1\}^{\{m+1,\ldots,n\}}} f(zy).$$

Here zy means the obvious concatenation of z and y. It is easy to check that f_K is monotone because f is. Next, it is an easy exercise to check that $\mathbf{I}(f_K) = \mathbf{I}_K(f)$. Next, $\mathbb{E}(fM_K) = \mathbb{E}(f_K M_K)$; to see this, note f_K is the conditional expectation of f onto the bits in K and M_K is measurable with respect to these bits.

Using the above and Lemma 12.48, we then obtain

$$\mathbf{I}_K(f) = \mathbf{I}(f_K) \le C \sqrt{|K|} \mathbb{E}(f_K M_K)\left(1 + \sqrt{-\log(\mathbb{E}(f_K M_K))}\right)$$

$$= C \sqrt{|K|} \mathbb{E}(fM_K)\left(1 + \sqrt{-\log(\mathbb{E}(fM_K))}\right).$$

□

Lemma 12.50 *If $c_1 \ge c_2 \ge \cdots \ge c_n > 0$, then*

$$\max\{\sum a_i^2 : a_1 \ge a_2 \ge \cdots \ge a_n \ge 0 : \forall k, \sum_{i=1}^k a_i \le \sum_{i=1}^k c_i\} = \sum c_i^2.$$

The proof is an exercise that is left to readers but we make the following comments. Existence of a maximum follows from compactness. The fact that the function x^2 is convex implies that

$$\max\{a^2 + b^2 : 0 \le a, b, a + b \le c\} = c^2.$$

This says that you should take one term as high as possible. This is "why" the result is true but the proof takes a few steps because one needs to worry about "boundary conditions."

Proof of Theorem 12.46. Assume without loss of generality that $I_1(f) \ge \cdots \ge I_n(f)$. Corollary 12.49 implies (using that $x(1 + \sqrt{\log(1/x)})$ is increasing for small x) that

$$\sum_{i=1}^k I_i(f) \le C \sqrt{k}(\Lambda(f))(1 + \sqrt{-\log(\Lambda(f))}).$$

Choose $c_1, \ldots, c_n$ so that for each k, we have

$$\sum_{i=1}^{k} c_i = C \sqrt{k}(\Lambda(f))(1 + \sqrt{-\log(\Lambda(f))})$$

Since $\sqrt{x}$ is concave, the c_i's are weakly decreasing. Lemma 12.50 then gives that

$$\mathbf{H}(f) \leq \sum_{k=1}^{n} C^2(\Lambda(f))^2(1 + \sqrt{-\log(\Lambda(f))})^2(\sqrt{k} - \sqrt{k-1})^2$$

$$\leq \sum_{k=1}^{n} C^2(\Lambda(f))^2(1 + \sqrt{-\log(\Lambda(f))})^2(1/k) \leq C_1(\Lambda(f))^2(1 - \log(\Lambda(f))) \log n.$$

$\square$

We end this section with the following theorem whose proof we omit. It states that for monotone functions, being noise sensitive is equivalent to being asymptotically uncorrelated with all *weighted* majority functions. Given $n \geq 1$, $w \in [0, 1]^n$ and $\omega = (x_1, \ldots, x_n) \in \{\pm 1\}^n$, let

$$M_{n,w}(\omega) := \text{sign}(\sum_{i=1}^{n} w_i x_i).$$

When the w_i's are in ± 1, we (essentially) obtain the majority functions that we had earlier. For $f \colon \{\pm 1\}^n \to \mathbb{R}$, let $\tilde{\Lambda}(f) := \max\{|\mathbb{E}(f M_{n,w})| : w \in [0, 1]^n\}$.

Theorem 12.51   *(i) The family of events $\{M_{n,w}(\omega) > s\}_{n \geq 1, w \in [0,1]^n, s \in \mathbb{R}}$ is uniformly stable (in the obvious sense).*
(ii) The sequence of monotone Boolean functions $\{f_n\}$ is noise sensitive if and only if $\lim_{n \to \infty} \tilde{\Lambda}(f_n) = 0$.

As far as (i) above, in (BKS99), it was shown that

$$\sup_{n \geq 1, w \in [0,1]^n, s \in \mathbb{R}} \mathbb{P}(M_{n,w}(\omega) > s, M_{n,w}(\omega_\epsilon) \leq s) \leq O(1)\epsilon^{1/4}.$$

In (P04), this upper bound was improved to $O(1)\epsilon^{1/2}$ which is necessarily the sharp exponent which one sees by looking at ordinary Majority and taking $n \to \infty$.

12.13 The BKS algorithmic noise sensitivity result

In the original proof of noise sensitivity of percolation crossing events in (BKS99), Theorem 12.46 was exploited together with an argument that involved an algorithm. This was done only for percolation but their argument proved the following more general result which we record and prove here.

Theorem 12.52 *Let $\{f_n\}$ be a sequence of monotone Boolean functions on $\{0, 1\}^{[n]}$ mapping into $\{0, 1\}$. Assume there is an integer B and constants C and δ so that the following holds. For all n, $[n]$ can be partitioned into at most B sets $A_1^n, A_2^n, \ldots, A_{k_n}^n$ (so $k_n \leq B$) so that for each $i = 1, \ldots, k_n$, there exists a randomized algorithm $\mathcal{A}^{n,i}$ that queries the bits in $[n]$, one bit at a time (the bit chosen may depend on the outcome of the values of the earlier bits) and then stops at some point such that*
(i) When $\mathcal{A}^{n,i}$ stops, $f_n(\omega)$ is determined and
(ii) For all $j \in A_i^n$, the probability that $\mathcal{A}^{n,i}$ queries bit j is at most C/n^δ.
Then $\{f_n\}$ is noise sensitive.

Before proving this, we note the following corollary.

Corollary 12.53 *If A_n is the event that there exists a crossing of an $n \times n$ square in two-dimensional critical bond percolation on the square lattice, then the sequence $\{A_n\}$ is noise sensitive.*

Proof We simply apply Theorem 12.52. We take $B = 2$. We take A_1^n to be the right-hand side of the box including the center line and take A_2^n to be the left-hand side of the box not including the center line.

We consider the following algorithm $\mathcal{A}^{n,1}$. Order all the edges arbitrarily. Let V_1 be the set of vertices on the left boundary. Choose the first (according to our arbitrary ordering) edge from V_1 to V_1^c and query that edge. If the edge is on, add the vertex of the edge which was in V_1^c to V_1. If not, don't. Continue looking at edges (in order) from V_1 to V_1^c that have not been checked before. (V_1 is then sort of a growing cluster.) Stop when we hit the right boundary (in which case, we know that there is a crossing) or when there are no further edges to check (in which case, we know that there is no crossing). The algorithm $\mathcal{A}^{n,2}$ is analogous but starts on the right boundary.

The algorithm $\mathcal{A}^{n,1}$ clearly determines the event in question. If j in is A_1^n, then if j is queried, there there is necessarily an open path from j to distance $n/2$ away. However, Theorem 2.1 and duality are easily known to imply that this latter event has probability bounded above by C/n^δ for some C and $\delta > 0$. Because the same holds for the algorithm $\mathcal{A}^{n,2}$, Theorem 12.52 allows us to conclude noise sensitivity. $\square$

Remark We will assume for simplicity that the algorithms are deterministic in that no exterior randomness is used; this was the case in the application of Theorem 12.52 to percolation crossings. The proof can be easily adapted to randomized algorithms. The main modifications in the proof is that x in the proof below should then be a function of ω, ω', z and the exterior randomness (rather than just a function of ω, ω', and z) and that when one conditions on ω, ω', one should also condition on the information of the exterior randomness that one has obtained at the completion of the randomized algorithm.

We first give the idea of the proof of the theorem. For $K \subseteq A_i^n$, the ith algorithm does not hit so many points in K and so f should be fairly uncorrelated with M_K, which implies by Lemma 12.48 that I_K is not so large; this is exactly the key lemma. Then, as in the proof of Theorem 12.46, we obtain $\Pi(f_n)$ is small yielding noise sensitivity.

The following lemma is key. We prove it later.

Lemma 12.54 *Assume the conditions of Theorem 12.52 (except we don't need to make the monotonicity assumption). There exists a constant C_1 so that for all n, for all $i = 1, \ldots, B$ and for all $K \subseteq A_i^n$, we have*

$$\mathbb{E}[f_n M_K] \leq C_1 (\log n)/n^{\delta/3}.$$

Proof of Theorem 12.52. The above lemma together with Corollary 12.49 (and a tiny computation) implies there is a constant C so that for all n, for all $i = 1, \ldots, B$ and for all $K \subseteq A_i^n$,

$$I_K(f_n) \leq C \sqrt{|K|}(\log n)^{3/2}/n^{\delta/3}.$$

Since we partition $[n]$ into at most B sets, we have that there exists a constant C_2 so that for all n and for all $K \subseteq [n]$, we have that

$$I_K(f_n) \leq C_2 \sqrt{|K|}(\log n)^{3/2}/n^{\delta/3}.$$

Now we use the proof method of Theorem 12.46. We assume without loss of generality that $I_i(f_n)$ is nonincreasing in i. We have from the last inequality that for each k

$$\sum_{j=1}^{k} I_j(f_n) \leq C_2 \sqrt{k}(\log n)^{3/2}/n^{\delta/3}.$$

As in the proof of Theorem 12.46, $\sum_{j=1}^{n} I_j(f_n)^2$ cannot be any larger than when equality holds in the above for all k. Hence

$$\Pi(f_n) \le \sum_{j=1}^{n} (C_2(\log n)^{3/2}/n^{\delta/3})^2((\sqrt{k} - \sqrt{k-1}))^2$$

$$\le C_3[(\log n)^3/n^{2\delta/3}]\log n = C_3(\log n)^4/n^{2\delta/3} \le C_4/n^{\delta/2}.$$

Now apply Theorem 1.21.   □

Before starting on the proof of Lemma 12.54, we state without proof two elementary probability facts without proof.

Lemma 12.55 *If $\{S_k\}$ is simple random walk, then for all m and a, we have that*

$$P(S_k \ge a \text{ for some } k \in \{1,\ldots,m\}) \le 2e^{-a^2/2m}$$

Lemma 12.56 *There exists a constant C so that if $\{S_k\}$ is simple random walk, then for all r and α, we have that*

$$P(|S_r| \le \alpha) \le C\alpha/\sqrt{r}$$

Proof of Theorem 12.54. Fix n, $i \in \{1,\ldots,B\}$, $K \subseteq A_i^n$ and consider the algorithm $\mathcal{A}^{n,i}$. Let ω, ω', and z be independent with ω uniform in $\{0,1\}^{[|K|]}$—; ω' uniform in $\{0,1\}^{[n-|K|]}$; and z uniform in $\{0,1\}^{[n]}$. Using these, we will choose a uniform configuration x from $\{0,1\}^{[n]}$ as follows. We run $\mathcal{A}^{n,i}$. When it chooses a bit in K to query, the value that we assign to that bit is the first bit of ω not yet used. When it chooses a bit not in K to query, the value that we assign to it is the first bit of ω' not yet used. Finally assign all bits not yet assigned using z. This final assignment is called x and it is clearly uniform. Since the algorithm determines f_n, we have that $f_n(x)$ is measurable with respect to ω and ω'.

Let V (for visited) be the random set of bits queried by $\mathcal{A}^{n,i}$. By assumption (ii) in Theorem 12.52, we easily obtain

$$E[|V \cap K|] \le |K|C/n^{\delta}.$$

Letting $A_1 := \{|V \cap K| \ge |K|/n^{2\delta/3}\}$, Markov's inequality yields that

$$P(A_1) \le C/n^{\delta/3}.$$

Next, let

$$A_2 := \{\exists j \in [1, |K|/n^{2\delta/3}] : |\sum_{i=1}^{j} \omega_i - j/2| \ge \sqrt{|K|/n^{2\delta/3}} \log n\}.$$

Lemma 12.55 and a computation yields that

$$P(A_2) \leq C/n^{\delta}.$$

Now, let

$$Q := \{|K \cap V| < |K|/n^{2\delta/3}\} \cap \{| \sum_{i=1}^{|K \cap V|} \omega_j - |K \cap V|/2| < \sqrt{|K|/n^{2\delta/3}} \log n\}.$$

Note Q is measurable with respect to ω, ω' and that $Q^c \subseteq A_1 \cup A_2$ and hence

$$P(Q^c) \leq C/n^{\delta/3}.$$

Now

$$|E[f_n M_K]| \leq |E[f_n I_{Q^c} M_K]| + |E[f_n I_Q M_K]|.$$

The first term is at most $P(Q^c) \leq C/n^{\delta/3}$. The second term is

$$|E[f_n I_Q M_K]| = |E[E[f_n I_Q M_K \mid \omega, \omega']]|$$

$$= |E[f_n E[I_Q M_K \mid \omega, \omega']]| \leq E[|E[I_Q M_K \mid \omega, \omega']|]$$

We claim $(\omega, \omega') \in Q$ implies that

$$|E[M_K \mid \omega, \omega']| \leq C_9 \log n/n^{\delta/3}. \tag{12.13}$$

This would then give us that $|E(f_n M_K)| \leq C \log n/n^{\delta/3}$, which is the desired result.

Note in the continuation that terms such as $|K \cap V|$ are now no longer random because we have conditioned on (ω, ω') with respect to which $|K \cap V|$ is measurable. Returning to prove (12.13), the reason this is true is essentially because M_K is affected by $(\omega, \omega') \in Q$ only if the sum of the other bits in K is closer to its mean than $\sqrt{|K|/n^{2\delta/3}} \log n$ but we make this more precise as follows. Note that $|\sum_{i=1}^{|K \cap V|} \omega_j - |K \cap V|/2| < \sqrt{|K|/n^{2\delta/3}} \log n$ implies that the difference (in absolute value) between the number of 1's and 0's in $K \cap V$ is at most $2\sqrt{|K|/n^{2\delta/3}} \log n$. Let W be the difference between the number of 1's and 0's in $K \setminus (K \cap V)$ and we let

$$U = \{|W| > 2\sqrt{|K|/n^{2\delta/3}} \log n\}.$$

Note that U is independent of (ω, ω') and from this, it is easy to see by symmetry that

$$E[M_K I_U \mid \omega, \omega'] = 0.$$

It follows that

$$|E[M_K \mid \omega, \omega']| = |E[M_K I_{U^c} \mid \omega, \omega']| \leq P(U^c) \tag{12.14}$$

where the independence of U and (ω, ω') is used again in the last inequality.

Now, using Lemma 12.56 for the first inequality below, we obtain

$$P(U^c) \le C \sqrt{|K|/n^{2\delta/3}} \log n \left(1 / \sqrt{|K \backslash (K \cap V)|}\right)$$

$$\le C \sqrt{|K|/n^{2\delta/3}} \log n \left(1 / \sqrt{|K|(1 - n^{-2\delta/3})}\right) \le C \log n / n^{\delta/3}$$

and hence by (12.14) that $|E[M_K \mid \omega, \omega']| \le C \log n / n^{\delta/3}$ when $(\omega, \omega') \in Q$ as desired.   □

Remarks As mentioned earlier, the disadvantages of the present approach compared to that of Section 8.2 of Chapter 8 is that monotonicity of the functions involved is necessary; the argument ultimately relies on hyper-contractivity; and it does not, in the case of percolation crossings, yield the polynomial quantitative noise sensitivity results. However, we feel that the argument is interesting and worth understanding. It would be interesting to have an example for which Theorem 12.52 could be applied but Theorem 8.2 could not be.

12.14 Black noise in the sense of Tsirelson

In the late 1990s, about at the same time as the seminal work on noise sensitivity (BKS99) of Benjamini, Kalai, and Schramm, Boris Tsirelson developed a deep theory of noises and in particular of **black noises**. A very good introduction on the topic can be found in Tsirelson's Saint-Flour lecture notes (Tsi04). Still, we wish to explain at a (very) informal level what this theory is about.

Informal definition of a noise in the one-dimensional setting

Consider a probability space $(\Omega, \mathcal{F}, \mathbb{P})$. A one-dimensional *noise* on this probability space is a two-parameter filtration $(\mathcal{F}_{s,t})_{s<t\in\mathbb{R}}$ that satisfies the following factorization property: for any $s < u < t$,

$$\mathcal{F}_{s,t} = \mathcal{F}_{s,u} \otimes \mathcal{F}_{u,t},$$

by which we mean that $\mathcal{F}_{s,u}$ is independent of $\mathcal{F}_{u,t}$ and the sub-σ-fields $\mathcal{F}_{s,u}$ and $\mathcal{F}_{u,t}$ together generate the larger one $\mathcal{F}_{s,t}$. There are two other technical constraints that need to be satisfied in order to be a proper noise: a certain translation invariance under time-shifts as well as a continuity property of the kind $\mathcal{F}_{s,t}$ is generated by $\bigcup_{\epsilon>0} \mathcal{F}_{s+\epsilon, t-\epsilon}$. We do not give more details on this and we refer to (Tsi04).

Noises in higher dimension

The above informal definition easily extends to the case of dimensions $d \geq$ 2, where one considers instead subfiltrations indexed by hyperrectangles $\mathcal{F}_{[s_1,t_1] \times \dots [s_d,t_d]}$.

This abstract setup developed by Tsirelson is fruitful in two respects:

(1) It is a convenient framework in order to consider scaling limits of discrete stochastic processes "ω^n". More precisely, one way to build interesting noises is to obtain them as limits of stochastic processes $\omega^n \in (\Omega^n, \mathcal{F}^n, (\mathcal{F}^n_{s,t})_{s<t}, \mathbb{P}^n)$, as $n \to \infty$. We give examples of this below.

(2) The factorization property $\mathcal{F}_{s,t} = \mathcal{F}_{s,u} \otimes \mathcal{F}_{u,t}$ enables a **spectral** study of the noise. More precisely, to each "observable" $f \in L^2(\Omega, \mathcal{F}, \mathbb{P})$, one can associate a spectral measure μ_f on the space of compact subsets of $\mathbb{R}$, which is obtained using the conditional expectations $\mathbb{E}[f \mid \mathcal{F}_{s,t}]$. See the Definition below.

Definition 12.57 (Spectral measure)

If one is given a noise $(\Omega, \mathcal{F}, (\mathcal{F}_{s,t})_{s<t}, \mathbb{P})$, then to each observable $f \in L^2(\Omega, \mathcal{F}, \mathbb{P})$, one can associate a spectral measure μ_f on the set of compact subsets of $\mathbb{R}$ characterized as in Proposition 9.5 by

$$\mu_f(\mathscr{S}_f \subset A) := \mathbb{E}[\mathbb{E}[f \mid \mathcal{F}_A]^2],$$

for any $A \subset \mathbb{R}$ made of finitely many intervals.

Let us now give some examples. We will remain vague on the way filtrations are obtained as scaling limits of discrete processes. See in particular Sections 2 and 3 in (Tsi04). The idea if that if ω_n is a certain discrete stochastic process, we choose a countable family of "observables" $f_n \in L^2(\Omega^n, \mathcal{F}^n, \mathbb{P}^n)$ which describe "well" what the stochastic process ω_n is and we pass to the limit $n \to \infty$ in order to obtain observables $f \in L^2(\Omega, \mathcal{F}, \mathbb{P})$ on a limiting object $\omega \in \Omega$. The limiting probability space and its induced filtration $\mathcal{F}_{s,t}$ depends a lot on the choice of observables f_n as we will see in the example of percolation below.

Example 12.58 Brownian motion $(B_t)_{t \in \mathbb{R}}$ with $B_0 = 0$ produces a one-dimensional noise by taking the filtration

$$\mathcal{F}_{s,t} := \sigma\{B_u - B_v, s < u < v < t\}$$

Example 12.59 (Coalescing flows or Brownian Web)   Consider the plane $\mathbb{R}^2$ as space $\times$ time, that is, $(x, t) \in \mathbb{R} \times \mathbb{R}$. Imagine one starts a Brownian motion at "each" point $(x, t) \in \mathbb{R} \times \mathbb{R}$, and one lets particles coalesce as in Figure 12.2 representing a discrete flow. Some work is needed to give a proper meaning to such a continuous coalescing flow and one possible way is to build a one-dimensional **coalescing noise** as a limit of systems of rescaled discrete coalescing random walks. (See (TW98) or (FN06) for another approach). See (Tsi04). A possible set of observables is to follow the trajectories of particles starting on $\mathbb{Q}^2$.

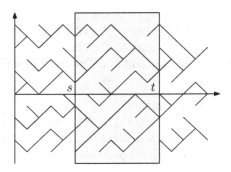

Figure 12.2 A discrete coalescing flow of random walks. The filtration $\mathcal{F}^n_{s,t}$ in some sense contains all the information concerning the coalescing flow between time s and time t.

Example 12.60

A very interesting example is given by the following 2-d "tabular" of black and white hexagons. There are two very distinct ways to take a scaling limit as the mesh of the hexagonal lattice goes to zero:

- One way is to build the filtration $\mathcal{F}^n$ using "linear" observables of the type

$$f^n_A := \frac{1}{n} \sum_{x_i \in \frac{1}{n} \mathbb{T} \cap A} x_i,$$

where A is a Borel subset of $[0,1]^2$ and $x_i \in \{-1,1\}$ depending on the color of the hexagons. To make this set of observables countable, one can restrict this family to the set of rectangles A with corners in $\mathbb{Q}^2$. With this point of view, by letting $n \to \infty$, one obtains at the limit the so-called two-dimensional **white noise**.

- On the other hand, if one builds the filtration $\mathcal{F}^n$ out of a set of countably many percolation crossing events $f_{n,Q}$ ranging over a dense set of quads $Q \subset \mathbb{R}^2$ (where a quad is a subset homeomorphic to a circle together with four distinguished points), then the limiting object is of a very different nature. We let $\mathcal{F}_S$ denote the σ-algebra generated by all the quads contained in S. It is easy to check that the independence property survives in the limit, namely that $\mathcal{F}_S$ is independent of $\mathcal{F}_{S'}$ if $S, S' \subset \mathbb{R}^2$ are (homeomorphic to) rectangles with disjoint interiors. As far as we are still on the discrete level, it is clear that $\mathcal{F}^n_{S \cup S'} = \mathcal{F}^n_S \otimes \mathcal{F}^n_{S'}$ but this factorization property is far from being obvious at the scaling limit $n \to \infty$. See the experimental paper (Tsi05) for an interesting discussion on this difficulty. In particular, Tsirelson conjectured in (Tsi05) that one still has $\mathcal{F}_{S \cup S'} = \mathcal{F}_S \otimes \mathcal{F}_{S'}$ and that the scaling limit of planar percolation should then be a noise. This factorization property was proved by Schramm and Smirnov in (SS11) thus showing that the scaling limit of planar percolation is a noise.

The reason why the noise of percolation is very different from the first point of view which leads to a 2-d white noise can be seen for example by showing that this noise of percolation is a **black noise** according to the following (informal) definition.

Definition 12.61 (black noise)
[We give the definition in the one-dimensional setting for simplicity.] A noise $(\Omega, \mathcal{F}, (\mathcal{F})_{s,t}, \mathbb{P})$ is **black** if and only if **all** its observables $f \in L^2(\Omega, \mathcal{F}, \mathbb{P})$ are noise sensitive. There are two ways to define noise sensitivity in this continuous setting.

- Either one resamples the "data" in small intervals of time $[k/L, (k+1)/L], k \in \mathbb{Z}$ each time with probability some fixed parameter $\epsilon > 0$ (see Section 5 in (Tsi04)) and we ask each functional $f \in L^2(\Omega, \mathcal{F}, \mathbb{P})$ to be noise sensitive under this noising procedure as $L \to \infty$. Note that the resampling procedure is possible thanks to the factorization property $\mathcal{F}_{k/L,(k+2)/L} = \mathcal{F}_{k/L,(k+1)/L} \otimes \mathcal{F}_{(k+1)/L,(k+2)/L}$.
- Another (equivalent) definition is to say that a functional $f \in L^2(\Omega, \mathcal{F}, \mathbb{P})$

is noise sensitive if and only if its spectral measure μ_f does not give any spectral mass to the set of finite nonempty subsets of $\mathbb{R}$.

Note the strong analogy with Proposition 4.2. Once the noise property of percolation is established (SS11), proving that it is a black noise is easier (see (Tsi04)). Another example of black noise is given by the above coalescing flow of Brownian motion (see (Tsi04) and (LJR04) for two different proofs).

12.15 Chaos, superconcentration, and multiple valleys

In (Ch08; Ch09), Sourav Chatterjee sheds some new light on the study of the fluctuations of a large variety of "Gaussian disordered models." He considers in particular the following models:

1. First passage percolation on $\mathbb{Z}^d$ for a general class of *edge-weights* (which is much larger than the law we considered in Chapter 7, where we had a Bernoulli $(1/2, 1/2)$ variable on $\{a, b\}$ with $0 < a < b < \infty$). See also the paper (BR08) for a thorough study of which laws on edges lead to the Benjamini, Kalai, Schramm bound of $|v|/\log|v|$.
2. The $(1 + 1)$-dimensional Gaussian random polymer model (i.e., in $\mathbb{Z}_+^2$). It was then generalized by Graham (Gr12) to the $(1 + d)$-dimensional case.
3. The largest eigenvalue of GUE matrices.
4. The Sherrington-Kirkpatrick model of spin glasses. In particular the fluctuation of the energy of its *ground state*.

In (Ch08), Chatterjee proves an equivalence property between three distinct behaviors:

1. **Superconcentration** (or anomalous fluctuations in our terms)
2. **Chaoticity** of the "minimizing state" (which would be a *geodesic* in the first example, a *polymer* in the second case, an *eigenvector* in the third example and finally a *ground state* in the last one), by which he means that the minimizing state is very sensitive to noise.
3. **Multiple valleys property**, by which he means that there are "almost minimizing states" that are very different (in fact "orthogonal") to the unique minimizing state.

As such, this work of Chatterjee highlights in a quantitative manner the link between anomalous fluctuations (or superconcentration) and noise sensitivity. Explaining his deep approach would take us too far and furthermore the book in progress (Ch14) on this topic will be a natural companion reading to this book.

13

Further directions and open problems

We end this book with a concise list of open questions which are grouped by theme.

13.1 Randomized algorithms

13.1.1 Best randomized algorithm for Iterated Majority

It may look surprising (owing to the simple iterative structure of the Iterated Majority function introduced in Example 1.5), but the following problem is to our knowledge not settled.

Open Problem 13.1 *Find the randomized algorithm with smallest possible revealment for the Iterated Majority function f_k on $n = 3^k$ bits. Even the order of magnitude of the decay to 0 (i.e., the exponent) is not known although there are some known bounds.*

13.1.2 Best randomized algorithms for critical percolation

Open Problem 13.2

(a) *Find algorithms for crossing events in percolation on the triangular lattice that examine on average at most $n^{7/4-\epsilon}$ sites for some $\epsilon > 0$.*

(b) *Show that any algorithm examines on average at least $n^{6/4+\epsilon}$ sites for some $\epsilon > 0$. (This would strengthen the lower bound obtained in Section 8.4 in Chapter 8).*

13.1.3 Randomized algorithm and spectrum

One way to obtain better bounds on the noise sensitivity of a Boolean function through the randomized algorithm approach (e.g., the percolation crossing events f_n) is to find the algorithms with the smallest possible revealments. Another possible direction is to address the following problem.

Open Problem 13.3 *Prove a sharper version of Theorem 8.2 from (SS10).*

13.1.4 Randomized algorithms and pivotal points

In (PSSW07), the authors introduce a seemingly very efficient randomized algorithm: one starts at some random initial site (using a well-chosen initial distribution) and then at each step, one picks the site that has the largest probability to be pivotal conditioned on what has been revealed so far. See (PSSW07). Their numerical simulations suggest that this randomized algorithm is more efficient than the one we used in Chapter 8 using an interface. Yet, nothing is rigorously known.

Conjecture 13.4 *Show that the algorithm from (PSSW07) that examines the bit that is most pivotal at the time examines on average at most $n^{7/4-\epsilon}$ sites for some $\epsilon > 0$. (This would then be more efficient than the randomized algorithm obtained in Theorem 8.4 from Chapter 8).*

In fact even the following weaker conjecture is unknown.

Conjecture 13.5 *Show that the algorithm from (PSSW07) mentioned above examines on average at most $n^{2-\epsilon}$ sites for some $\epsilon > 0$.*

Observe that in general, it perhaps is not always an optimal strategy to choose the most pivotal point when choosing the next variable.

13.1.5 Proving the existence of exceptional times on $\mathbb{Z}^2$ using randomized algorithms

So far the only proof of Theorem 11.8 that yields the existence of exceptional times on $\mathbb{Z}^2$ is the proof from (GPS10) which relies on the geometric approach highlighted in Chapter 10. It would be very appealing to have a much shorter proof of this fact using the randomized algorithm approach from (SS10).

Open Problem 13.6 *Prove the existence of exceptional times on the square lattice $\mathbb{Z}^2$ (i.e., Theorem 11.8) using randomized algorithms.*

It is easy to see that proving this open problem boils down to proving that there exists $C > 0$ and $\epsilon > 0$ such that the following holds for all $r \geq 1$:

$$\alpha_2(r) \leq C r^{-\epsilon} \alpha_1(r)^2. \tag{13.1}$$

Note that this relation is in the same spirit as (11.15) obtained by Beffara (see the appendix in (GPS10)). However, to our knowledge, the above one is not proven yet.

13.1.6 Do small witnesses imply noise sensitivity?

Consider a random witness W (see Definition 8.11) for a Boolean function f on Ω_n, where the randomness may depend on external randomness.

Definition 13.1   The **revealment of a witness** W for a Boolean function f, denoted by δ_W, is defined to be

$$\delta_W := \max_{i \in [n]} \mathbb{P}(i \in W).$$

The **witness revealment of a Boolean function** f, denoted by δ_f^{Wit}, is defined by

$$\delta_f^{\mathrm{Wit}} := \inf_W \delta_W$$

where the infimum is taken over all randomized witnesses W for f.

The following question, if answered positively, would strengthen the phenomenon highlighted in Corollary 8.3.

Open Problem 13.7   *If $\{f_n\}$ is a sequence of Boolean functions such that*

$$\lim_{n \to \infty} \delta_{f_n}^{\mathrm{Wit}} = 0,$$

does it follow that $\{f_n\}$ is noise sensitive?

13.2 Dynamical percolation

We collect here some of the main open problems that remain for dynamical percolation on $\mathbb{T}$ or $\mathbb{Z}^2$. The main obstruction to obtain sharp results in the case of the triangular lattice $\mathbb{T}$ comes form the fact that the geometric approach developed in (GPS10) (see also Chapter 10) works so far only for **monotone** events.

Conjecture 13.8   *For dynamical percolation on the triangular lattice $\mathbb{T}$, prove*

(a) *That the Hausdorff dimension of the exceptional times for which there is both a white and a black infinite cluster is a.s. $2/3$. (The proof of the lower bound $1/9$ was achieved in (GPS10)).*

(b) That there exist exceptional times for dynamical percolation for which there are two infinite arms and one infinite dual arm and show that the Hausdorff dimension of this set of times is a.s. 1/9.

As mentioned previously, this conjecture would follow form the following problem.

Open Problem 13.9 *Extend the geometric approach of the Fourier spectral sample $\mathscr{S}_{f_n}$ from (GPS10) to the case of non-monotone functions f_n.*

Let us end this section with the following problem.

Open Problem 13.10 *Prove that there exist exceptional times for dynamical percolation on the square lattice $\mathbb{Z}^2$ for which there is both a white and a black infinite cluster.*

13.3 The spectral sample vs. the pivotal set

In what follows, let f_n be once again the left–right crossing event of a n by n square for critical percolation on the triangular lattice $\mathbb{T}$. In (GPS10), it is proved that if one considers the rescaled spectral sample $\frac{1}{n}\mathscr{S}_{f_n}$ as a random compact subset of the unit square $[0,1]^2$, then this random set has a scaling limit as $n \to \infty$ that we shall denote by $\mathscr{S}_\infty \subset [0,1]^2$. Similarly, one may consider the rescaled pivotal set $\frac{1}{n}\mathcal{P}_{f_n}$ which is known to converge as $n \to \infty$ to a random compact subset $\mathcal{P}_\infty$ of the unit square.

It is not hard to see that $\mathcal{P}_\infty$ and $\mathscr{S}_\infty$ do not have the same law (their "three-point" function differ). It is conjectured in (GPS10) that they look very different in the following sense.

Conjecture 13.11 *The laws of $\mathscr{S}_\infty$ and $\mathcal{P}_\infty$ are mutually singular.*

Proving such a conjecture would in some sense highlight the fact that there is no hope to understand **noise sensitivity** simply by studying the pivotal set. See also Section 12.2 in the miscellaneous Chapter 12.

13.4 Noise sensitivity and exceptional times away from the independent case

13.4.1 Conservative dynamics (exclusion process) on percolation

In (BGS13), noise sensitivity for percolation under a large class of symmetric exclusion process dynamics was proved. (See Subsection 12.6.2 in

Chapter 12.) However, the proof in (BGS13) requires the exclusion kernel P to be symmetric and "medium-range" in the sense that there is an exponent $a > 0$ such that

$$[P(x,y)]_{x,y \in \mathbb{T}} \asymp \frac{1}{|x-y|^{2+a}}.$$

(The exponent 2 is there to make the kernel integrable.) In particular the techniques used in (BGS13) are far from being able to handle the most interesting case, which is the nearest neighbor simple exclusion process.

Open Problem 13.12   *Prove that percolation is noise sensitive under the nearest neighbor simple exclusion process on $\mathbb{Z}^2$ or $\mathbb{T}$.*

So far nothing has been proved concerning the existence of exceptional times for dynamical percolation under such exclusion processes:

Open Problem 13.13

- *Prove that there are exceptional times on $\mathbb{T}$ for the above medium-range exclusion dynamics with exponent $a > 0$ (even the case where a is very small, corresponding to the heavy-tailed case, is not settled).*
- *Prove the existence of exceptional times on $\mathbb{T}$ under the nearest-neighbor exclusion process (this should be much harder).*

Finally the aim of the next open question is to highlight the fact that the property of having exceptional times under a conservative dynamics should be stronger than the same phenomenon for ordinary dynamical percolation.

Open Problem 13.14   *Is it the case that if some event has exceptional times for an exclusion process, then it necessarily has exceptional times for ordinary dynamical percolation?*

13.4.2 Random-cluster model

The random cluster model is a dependent percolation model (also called FK percolation) that has been successfully introduced to study the correlation structure of various ferromagnetic models (including the Ising model discussed below). On a finite graph $G = (V, E)$, each configuration $\omega \in \{0, 1\}^E$ has a probability proportional to $p^{|\omega|}(1-p)^{|E|-|\omega|}q^{k(\omega)}$, where $|\omega|$ is the number of 1's (or open edges) and $k(\omega)$ is the number of connected components in ω. The dependency structure comes from the factor $q^{k(\omega)}$ (unless of course $q = 1$ which corresponds to standard i.i.d. bond percolation). See the book (Gri06). One can define a natural heat-bath dynamics $(\omega_t)_{t \geq 0}$ on

FK configurations which preserves the random-cluster measure (see also (Gri06)). As opposed to the Glauber dynamics discussed below, this heat-bath dynamics is not local.

Open Problem 13.15 *Consider* $(\omega_t)_{t \geq 0}$ *a heat-bath dynamics for the* $q = 2$ *random cluster model on* $\mathbb{Z}^2$ *at the critical temperature* $p_c(2) = \frac{\sqrt{2}}{1+\sqrt{2}}$. *Prove the following:*

(a) *Large crossing events are noise sensitive under the dynamics* (ω_t).
(b) *There exist exceptional times for which there is an infinite cluster in* ω_t.

See (DGP14) for a more detailed discussion as well as a precise conjecture. For example it is conjectured in (DGP14) that the Hausdorff dimension of the set of exceptional times for $q = 2$ is a.s. $10/13$. It is also conjectured that for all $q \in (q^*, 4]$, with $q^* = 4\cos^2(\frac{\pi}{4}\sqrt{14}) \approx 3.83$ there are a.s. no exceptional times for the heat-bath dynamics for FK percolation with parameter q (at the corresponding critical value $p_c(q)$ for p) even though there are pivotal points at all scales! If true, this would highlight a rather counterintuitive phenomenon.

13.5 Glauber dynamics and Ising model

In this section, we consider the Ising model on $\mathbb{Z}^2$ at critical inverse temperature $\beta = \beta_c$ (see, e.g., (Gri06) for background) on the $n \times n$ square Λ_n with either $+$ or *free* boundary conditions. The so-called **Glauber dynamics** is a very simple and celebrated dynamics which preserves the Ising measure $\mathbb{P}_{\beta_c}$. It induces a dynamics $\sigma_t \in \{\pm\}^{\Lambda_n}$ where *spins* are updated at rate one depending on the values of the spins nearby. Preserving the measure $\mathbb{P}_{\beta_c}$ means that if one starts at equilibrium (i.e., $\sigma_0 \sim \mathbb{P}_{\beta_c}$), then for any $t > 0$, $\sigma_t \sim \mathbb{P}_{\beta_c}$.

Open Problem 13.16 *Prove that crossing events are noise stable under the Glauber dynamics. More precisely, if* f_n *is the* $\{\pm 1\}$-*event that there is a* $+$ *cluster crossing form left to right in* Λ_n *(it can be shown using (DHN11) that the variance of* f_n *is non-degenerate at* $\beta = \beta_c$) *then for any* $t > 0$, *one has*

$$\text{Cov}(f_n(\sigma_0), f_n(\sigma_t)) \to 1,$$

as $n \to \infty$ *where* $\sigma_t \in \{\pm\}^{\Lambda_n}$ *is an instance of a Glauber dynamics starting at equilibrium. If true, this would mean that the spin-clusters of the Ising model evolve very slowly under the Glauber dynamics.*

A consequence of this stability property would enable to address the following problem

Open Problem 13.17 *Consider Glauber dynamics at criticality for the Ising model on $\mathbb{Z}^2$. Prove that a.s. there are no exceptional times with an infinite + cluster.*

13.6 Deterministic and randomized complexity

Open Problem 13.18 *(The generalized Aanderaa–Rosenberg Conjecture due to Rivest and Vuillemin) Is Theorem 12.32 true if we drop the assumption that n is a prime power?*

Remark The conjecture is in fact stronger than this in that the assumptions of monotonicity and nontriviality of the property are replaced by the weaker assumption that the function values at the all 0 configuration and the all 1 configuration differ. Theorem 12.32 is in fact proved under this weaker assumption (together with the assumptions of transitivity and n being a prime power).

Open Problem 13.19 *(Karp Conjecture) Does there exist $c > 0$ so that for all n and for all nontrivial monotone graph properties on n vertices, the corresponding randomized complexity (see Definition 12.33) is at least cn^2?*

13.7 A phase transition in the k-SAT problem

Recall the following conjecture from Section 12.5 in Chapter 12:

Conjecture 13.20 *Friedgut proved a sharp threshold property for k-SAT satisfiability in (Fri99) around $M = c_k(n)n$, where $c_k(n)$ is bounded away form 0 and ∞ as $n \to \infty$.*
Prove that $c_k(n)$ converges to some limiting value as $n \to \infty$. (In other words, show that the location of the phase transition stabilizes as $n \to \infty$). This is the same thing as proving that $c_k = c_k^*$ as stated in Conjecture 12.1.

As explained in Section 12.5 of Chapter 12, there are some very interesting (nonrigorous) works on this problem based on the so-called *cavity method*. See (MMZ06) and references therein. Let us also point out the recent paper (BGT13) which, based on statistical physics ideas, gives strong rigorous indication that the above sharp-threshold indeed converges.

13.8 Anomalous fluctuations

Open Problem 13.21 *Let $0 < a < b$ be fixed and consider the first passage percolation model on $\mathbb{Z}^2$ from Chapter 7.*

(a) *Prove that that there exists $\epsilon > 0$ such that (using the notations from Chapter 7)*

$$\text{Var}[\text{dist}_\omega(0, v)] \le |v|^{1-\epsilon}, .$$

Such an upper bound would already greatly improve on the upper bound $O(n/\log n)$ obtained in (BKS03) and in our Theorem 7.3.

(b) *Prove the conjectured fluctuation of order $n^{1/3}$, that is, that*

$$\text{Var}[\text{dist}_\omega(0, v)] = O(|v|^{2/3})$$

(c) *Prove the existence of a limiting law for the fluctuations. Namely prove that*

$$\frac{\text{dist}_\omega(0, v) - \lambda|v|}{|v|^{1/3}}$$

converges in law toward a limiting distribution where λ is an appropriate constant that depends on the direction of the point v. Finally relate this law with the celebrated Tracy–Widom distribution.

Remark (1) Even though the techniques from Chapter 7 work for all dimensions $d \ge 2$, (b) and (c) above should hold only for $d = 2$.
(2) Note, as we discussed along Chapter 7, that this open problem has been settled by K. Johansson in his breakthrough paper (Joh00) for the related model of **directed last passage direction** where the weights on the edges are either geometric or exponential.

Another big open problem in the area is the following *universality* question.

Open Problem 13.22 *Prove that the law of the fluctuations in first passage percolation does not depend on the "microscopic structure" given by the underlying graph as well as the law of the i.i.d. weights on edges.*

Besides Theorem 7.3 from (BKS03) and its generalization to a large variety of weight distributions in (BR08), it seems one is still very far form answering the above two open problems in the case of FPP. Let us point out that there has been interesting progress in this direction recently by Sourav Chatterjee, who proved in (Ch13a) a theorem concerning relations between critical exponents in FPP (assuming they exist).

References

[AP04] Dimitris Achlioptas and Yuval Peres. The threshold for random k-SAT is 2k(ln2-O(k)). *J. Amer. Math. Soc.*, 17(4): 947–973, 2004.

[ABGM13] Daniel Ahlberg, Erik I. Broman, Simon Griffiths, and Robert Morris. Noise sensitivity in continuum percolation, *Israel J. Math.*, to appear.

[AS00] Noga Alon and Joel H. Spencer. *The Probabilistic Method*. Wiley-Interscience Series in Discrete Mathematics and Optimization. Wiley-Interscience [John Wiley & Sons], New York, second edition, 2000. With an appendix on the life and work of Paul Erdős.

[B65] John F. Banzhaf. Weighted voting doesn't work: A mathematical analysis *Rutgers Law Review*, 19(2):317343, 1965.

[BA91] David J. Barsky and Michael Aizenman. Percolation critical exponents under the triangle condition. *Ann. Probab.*, 19(4):1520–1536, 1991.

[BGT13] Mohsen Bayati, David Gamarnik, and Prasad Tetali. Combinatorial approach to the interpolation method and scaling limits in sparse random graphs. *Ann. Probab.*, 41(6):4080–4115, 2013.

[B75] William Beckner. Inequalities in Fourier analysis. *Ann. Math.*, 2nd series 102(1):159–182, 1975.

[B12] Vincent Beffara. Schramm-Loewner evolution and other conformally invariant objects "Probability and statistical physics in two and more dimensions." (D. Ellwood, C. Newman, V. Sidoravicius, and W. Werner, editors). *Proceedings of the Clay Mathematics Institute Summer School and XIV Brazilian School of Probability* (Buzios, Brazil), Clay Mathematics Proceedings 15 (2012), 1–48.

[BDC12] Vincent Beffara and Hugo Duminil-Copin. The self-dual point of the two-dimensional random-cluster model is critical for $q \geq 1$, *Prob. Theory Relat. Fields*, 153(3-4):511–542, 2012.

[BCHs+96] Mihir Bellare, Don Coppersmith, Johan Håstad, Marcos Kivi, and Madhu Sudan. Linear testing in characteristic two. *IEEE Trans. Informat. Theory*, 42(6):1781–1796, 1996.

[BOL87] Michael Ben-Or and Nathan Linial. Collective coin flipping. *Randomness and Computation* (S. Micali, ed.), Advances in Computing Research, Vol. 5. JAI Press, December 1989.

[BR08] Michel Benaïm and Raphaël Rossignol. Exponential concentration for first passage percolation through modified Poincaré inequalities. *Ann. Inst. Henri Poincaré Probab. Stat.*, 44(3):544–573, 2008.

[BKS99] Itai Benjamini, Gil Kalai, and Oded Schramm. Noise sensitivity of Boolean functions and applications to percolation. *Inst. Hautes Études Sci. Publ. Math.*, (90):5–43 (2001), 1999.

[BKS03] Itai Benjamini, Gil Kalai, and Oded Schramm. First passage percolation has sublinear distance variance. *Ann. Probab.*, 31(4):1970–1978, 2003.

[BSW05] Itai Benjamini, Oded Schramm, and David B. Wilson. Balanced Boolean functions that can be evaluated so that every input bit is unlikely to be read. In *STOC'05: Proceedings of the 37th Annual ACM Symposium on Theory of Computing*, pp. 244–250. ACM, New York, 2005.

[BLR93] Manuel Blum, Michael Luby, and Ronitt Rubinfeld. Self-testing/correcting with applications to numerical problems. In *Proceedings of the 22nd Annual ACM Symposium on Theory of Computing* (Baltimore, MD, 1990), pp. 549–595, 1993.

[BR06a] Béla Bollobás and Oliver Riordan. A short proof of the Harris-Kesten theorem. *Bull. London Math. Soc.*, 38(3):470–484, 2006.

[BR06b] Béla Bollobás and Oliver Riordan. Sharp thresholds and percolation in the plane. *Random Struct. Algorith.*, 29(4): 524–548, 2006.

[B70] Aline Bonami. Étude des coefficients de Fourier des fonctions de Lp(G). *Ann. Inst. Fourier (Grenoble)*, 20 fasc. 2, 335–402, 1970.

[Bor82] Christer Borell. Positivity improving operators and hypercontractivity. *Math. Z.*, 180(2):225–234, 1982.

[Bor85] Christer Borell. Geometric bounds on the Ornstein-Uhlenbeck velocity process. *Z. Wahrsch. Verw. Gebiete*, 70(1):1–13, 1985.

[Bou02] Jean Bourgain. On the distributions of the Fourier spectrum of Boolean functions. *Israel J. Math.*, 131:269–276, 2002.

[BKK+92] Jean Bourgain, Jeff Kahn, Gil Kalai, Yitzhak Katznelson, and Nathan Linial. The influence of variables in product spaces. *Israel J. Math.*, 77(1–2):55–64, 1992.

[BGS13] Erik I. Broman, Christophe Garban, and Jeffrey E. Steif. Exclusion sensitivity of Boolean functions. *Probab. Theor. Rel. Fields*, 155(3-4):621–663, 2013.

[Ch08] Sourav Chatterjee. Chaos, concentration, and multiple valleys. Preprint. arXiv:0810.4221, 2008.

[Ch09] Sourav Chatterjee. Disorder chaos and multiple valleys in spin glasses. Preprint. arXiv:0907.3381, 2009.

[Ch13a] Sourav Chatterjee. The universal relation between scaling exponents in first-passage percolation. To appear in *Annals Math.*, 177(2):663–697, 2013.

[Ch14] Sourav Chatterjee. Superconcentration and related topics. Springer Monographs in Mathematics. Springer, New York.

[DGP14] Hugo Duminil-Copin, Christophe Garban, and Gábor Pete. The near-critical planar FK-Ising model. *Commun. Math. Phys.*, 326(1):1-35, 2014.

[DHN11] Hugo Duminil-Copin, Clément Hongler, and Pierre Nolin. Pierre connection probabilities and RSW-type bounds for the two-dimensional FK Ising model. *Comm. Pure Appl. Math.* 64(9): 1165–1198, 2011.

[Feder69] Paul Federbush. Partially alternate derivation of a result of Nelson. *J. Math. Phys.*, 10: 50–52, 1969.

[FW07]   Klaus Fleischmann and Vitali Wachtel. Lower deviation probabilities for supercritical Galton-Watson processes. *Ann. Inst. Henri Poincaré Probab. Stat.*, 43(2): 233–255, 2007.

[FN06]   Luiz Renato Fontes and Charles Newman. The full Brownian web as scaling limit of stochastic flows. *Stoch. Dyn.*, 6(2): 213–228, 2006.

[Fri98]   Ehud Friedgut. Boolean functions with low average sensitivity depend on few coordinates. *Combinatorica*, 18(1):27–35, 1998.

[Fri99]   Ehud Friedgut. Sharp thresholds of graph properties, and the k-sat problem. With an appendix by Jean Bourgain. *J. Am. Math. Soc.*, 12(4): 1017–1054, 1999.

[Fri04]   Ehud Friedgut. Influences in product spaces: KKL and BKKKL revisited. *Combin. Probab. Comput.*, 13(1):17–29, 2004.

[FKW02]   Ehud Friedgut, Jeff Kahn, and Avi Wigderson. Computing graph properties of randomized subcube partitions. In *Randomization and Approximation Techniques in Computer Science*, vol. 2483 of Lecture Notes in Computes Science, pp. 105–113. Springer, Berlin, 2002.

[FK96]   Ehud Friedgut and Gil Kalai. Every monotone graph property has a sharp threshold. *Proc. Am. Math. Soc.*, 124(10):2993–3002, 1996.

[Gar11]   Christophe Garban. Oded Schramm's contributions to noise sensitivity. *Ann. Probab.*, 39(5):1702–1767, 2011.

[GPS10]   Christophe Garban, Gábor Pete, and Oded Schramm. The Fourier spectrum of critical percolation. *Acta Mathematica*, 205(1):19–104, 2010.

[Gr12]   Ben T. Graham. Sublinear variance for directed last-passage percolation. *J. Theor. Probab.* 25(3): 687–702, 2012.

[GG06]   Ben T. Graham and Geoffrey R. Grimmett. Influence and sharp-threshold theorems for monotonic measures. *Ann. Probab.*, 34(5):1726–1745, 2006.

[GG11]   Ben T. Graham and Geoffrey R. Grimmett. Sharp thresholds for the random-cluster and Ising models. *Ann. Appl. Probab.*, 21(1):240–265, 2011.

[Gri99]   Geoffrey Grimmett. *Percolation*, 2nd ed. Grundlehren der mathematischen Wissenschaften 321. Springer-Verlag, Berlin, 1999.

[Gri06]   Geoffrey Grimmett. *The Random-Cluster Model*. Grundlehren der Mathematischen Wissenschaften 333. Springer-Verlag, Berlin, 2006.

[Gro75]   Leonard Gross. Logarithmic Sobolev inequalities. *Am. J. Math.*, 97(4):1061–1083, 1975.

[HPS97]   Olle Häggström, Yuval Peres, and Jeffrey E. Steif. Dynamical percolation. *Ann. Inst. H. Poincaré Probab. Statist.*, 33(4):497–528, 1997.

[Haj91]   Péter Hajnal. An $\Omega(n^{4/3})$ lower bound on the randomized complexity of graph properties. *Combinatorica*, 11(2):131–143, 1991.

[Haj92]   Péter Hajnal. Decision tree complexity of Boolean functions. In *Sets, Graphs and Numbers* (Budapest, 1991), Vol. 60 of Colloq. Math. Soc. János Bolyai, pp. 375–389. North-Holland, Amsterdam, 1992.

[HPS13]   Alan Hammond, Gábor Pete, and Oded Schramm. Local time on the exceptional set of dynamical percolation, and the Incipient Infinite Cluster. Preprint. arXiv:1208.3826, 2012.

[HS94]   Takashi Hara and Gordon Slade. Mean-field behaviour and the lace expansion. In *Probability and Phase Transition* (Cambridge, 1993),

Vol. 420 of NATO Adv. Sci. Inst. Ser. C Math. Phys. Sci., pp. 87–122. Kluwer Academic, Dordrecht, 1994.

[LJR04] Yves Le Jan and Olivier Raimond. Sticky flows on the circle and their noises. *Probab. Theory Relat. Fields*, 129(1): 63–82, 2004.

[Joh00] Kurt Johansson. Shape fluctuations and random matrices. *Comm. Math. Phys.*, 209(2):437–476, 2000.

[KKL88] Jeff Kahn, Gil Kalai, and Nathan Linial. The influence of variables on boolean functions. *29th Annual Symposium on Foundations of Computer Science*, pp. 68–80, 1988.

[Kal02] Gil Kalai. A Fourier-theoretic perspective on the Condorcet paradox and Arrow's theorem. *Adv. Appl. Math.*, 29(3): 412–426, 2002.

[KK13] Nathan Keller and Guy Kindler. Quantitative relation between noise sensitivity and influences. *Combinatorica*, 33(1):45–71, 2013.

[Kes80] Harry Kesten. The critical probability of bond percolation on the square lattice equals $\frac{1}{2}$. *Commun. Math. Phys.*, 74(1):41–59, 1980.

[Kes87] Harry Kesten. Scaling relations for 2D-percolation. *Commun. Math. Phys.*, 109(1):109–156, 1987.

[KZ87] Harry Kesten and Yu Zhang. Strict inequalities for some critical exponents in two-dimensional percolation. *J. Statist. Phys.*, 46(5-6):1031–1055, 1987.

[KKMO07] Subhash Khot, Guy Kindler, Elchanan Mossel, and Ryan O'Donnell. Optimal inapproximability results for MAX-CUT and other 2-variable CSPs? *SIAM J. Comput.*, 37(1):319–357, 2007.

[Kor03] Alekseĭ D. Korshunov. Monotone Boolean functions. *Uspekhi Mat. Nauk*, 58(5(353)):89–162, 2003.

[LSW02] Gregory F. Lawler, Oded Schramm, and Wendelin Werner. One-arm exponent for critical 2D percolation. *Electron. J. Probab.*, 7(2), 13 pp. (electronic), 2002.

[LS] Eyal Lubetzky and Jeffrey E. Steif. Strong noise sensitivity and random graphs. *Ann. Probab.*, to appear.

[Lyo11] Russell Lyons. *Probability on Trees and Networks*. Cambridge University Press, in preparation. Written with the assistance of Y. Peres. In preparation. Current version available at http://php.indiana.edu/~rdlyons/.

[Mar74] Grigoriĭ A. Margulis. Probabilistic characteristics of graphs with large connectivity. *Problemy Peredači Informacii*, 10(2):101–108, 1974.

[MMZ06] Stephan Mertens, Marc Mézard abd Riccardo Zecchina. Threshold values of random K-SAT from the cavity method. *Random Struct. Algorith.*, 28(3): 340–373, 2006.

[Mez03] Marc Mézard. Optimization and physics: On the satisfiability of random Boolean formulae. (English summary). *Ann. Henri Poincaré*, 4 (Suppl. 1): S475–S488, 2003.

[MO03] Elchanan Mossel and Ryan O'Donnell. On the noise sensitivity of monotone functions. *Random Struct. Algorithms.*, 23(3):333–350, 2003.

[MO05] Elchanan Mossel and Ryan O'Donnell. Coin flipping from a cosmic source: On error correction of truly random bits. *Random Struct. Algorith.*, 26(4):418–436, 2005.

[MOO10] Elchanan Mossel, Ryan O'Donnell, and Krzysztof Oleszkiewicz. Noise stability of functions with low influences: invariance and optimality. *Ann. Math.*, 171(1):295–341, 2010.

[MOR+06] Elchanan Mossel, Ryan O'Donnell, Oded Regev, Jeffrey E. Steif, and Benny Sudakov. Non-interactive correlation distillation, inhomogeneous Markov chains, and the reverse Bonami-Beckner inequality. *Israel J. Math.*, 154:299–336, 2006.

[MOS03] Elchanan Mossel, Ryan O'Donnell, and Rocco P. Servedio. Learning juntas. In *STOC'03: Proceedings of the 35th Annual ACM Symposium on Theory of Computing*, pp. 206–212, ACM, New York, 2003.

[Nel66] Edward Nelson. A quartic interaction in two dimensions. In *Mathematical Theory of Elementary Particles (Proc. Conf., Dedham, MA, 1965)*, pp. 69–73. MIT Press, Cambridge, MA, 1966.

[NP95] Charles M. Newman and Marcelo S. T. Piza. Divergence of shape fluctuations in two dimensions. *Ann. Probab.*, 23(3):977–1005, 1995.

[Nol09] Pierre Nolin. Near-critical percolation in two dimensions. *Electron. J. Probab.*, 13(55):1562—1623, 2009.

[O'D] Ryan O'Donnell. History of the hypercontractivity theorem. http://boolean-analysis.blogspot.com/.

[O'D03a] Ryan O'Donnell. http://www.cs.cmu.edu/~odonnell/boolean-analysis/, 2003. Course homepage.

[O'D03b] Ryan O'Donnell. *Computational Applications of Noise Sensitivity*. PhD thesis, MIT, 2003.

[O'D08] Ryan O'Donnell. Some topics in analysis of Boolean functions. In *STOC'08*, pp. 569–578. ACM, New York, 2008.

[O'D14] Ryan O'Donnell. *Analysis of Boolean Functions*. Cambridge University Press, 2014.

[OSSS05] Ryan O'Donnell, Michael Saks, Oded Schramm, and Rocco Servedio. Every decision tree has an influential variable. *FOCS*, 2005.

[OS07] Ryan O'Donnell and Rocco A. Servedio. Learning monotone decision trees in polynomial time. *SIAM J. Comput.*, 37(3):827–844 (electronic), 2007.

[PP94] Robin Pemantle and Yuval Peres. Planar first-passage percolation times are not tight. In *Probability and Phase Transition* (Cambridge, 1993), Vol. 420 of *NATO Adv. Sci. Inst. Ser. C Math. Phys. Sci.*, pp. 261–264. Kluwer Academic, Dordrecht, 1994.

[P46] Lionel Penrose. The elementary statistics of majority voting *J. Roy. Statist. Soc.*, 109(1):5357, 1946.

[P04] Yuval Peres. Noise stability of weighted majority. arXiv:0412.5377, 2004

[PSSW07] Yuval Peres, Oded Schramm, Scott Sheffield, and David B. Wilson. Random-turn hex and other selection games. *Am. Math. Mon.*, 114(5):373–387, 2007.

[Re00] David Reimer. Proof of the van den Berg-Kesten conjecture. *Combin. Probab. Comput.*, 9(1):27–32, 2000.

[RW10] Oliver Riordan and Nicholas Wormald. The diameter of sparse random graphs. *Combin. Probab. Comput.*, 19(5-6):835–926, 2010.

[RV75] Ronald L. Rivest and Jean Vuillemin. A generalization and proof of the

Aanderaa-Rosenberg conjecture. *In Seventh Annual ACM Symposium on Theory of Computing* (Albuquerque, NM, 1975), pp. 6–11. Assoc. Comput. Mach., New York, 1975.

[Ros08] Raphaël Rossignol. Threshold phenomena on product spaces: BKKKL revisited (once more). *Electron. Comm. Probab.*, 13, 35–44, 2008.

[Rus81] Lucio Russo. On the critical percolation probabilities. *Z. Wahrsch. Verw. Gebiete*, 56(2):229–237, 1981.

[Rus82] Lucio Russo. An approximate zero-one law. *Z. Wahrsch. Verw. Gebiete*, 61(1):129–139, 1982.

[Sch00] Oded Schramm. Scaling limits of loop-erased random walks and uniform spanning trees. *Israel J. Math.*, 118:221–288, 2000.

[SS11] Oded Schramm and Stanislav Smirnov. On the scaling limits of planar percolation. With an appendix by Christophe Garban. *Ann. Probab., volume in honor of Oded Schramm*, 39(5):1768–1814, 2011.

[SS10] Oded Schramm and Jeffrey Steif. Quantitative noise sensitivity and exceptional times for percolation. *Ann. Math.*, 171(2):619–672, 2010.

[Smi01] Stanislav Smirnov. Critical percolation in the plane: Conformal invariance, Cardy's formula, scaling limits. *C. R. Acad. Sci. Paris Sér. I Math.*, 333(3):239–244, 2001.

[SW01] Stanislav Smirnov and Wendelin Werner. Critical exponents for two-dimensional percolation. *Math. Res. Lett.*, 8(5-6):729–744, 2001.

[Ste09] Jeffrey Steif. A survey of dynamical percolation. *Fractal Geometry and Stochastics*, IV, Birkhauser, Boston, pp. 145–174, 2009.

[Tal94] Michel Talagrand. On Russo's approximate zero-one law. *Ann. Probab.*, 22(3):1576–1587, 1994.

[Tal96] Michel Talagrand. How much are increasing sets positively correlated? *Combinatorica*, 16(2):243–258, 1996.

[Tal97] Michel Talagrand. On boundaries and influences. *Combinatorica*, 17(2):275–285, 1997.

[TW98] Balint Toth and Wendelin Werner. The true self-repelling motion. *Probab. Theory Relat. Fields*, 111(3): 375–452, 1998.

[Tsi04] Boris Tsirelson. Scaling limit, noise, stability. Lectures on probability theory and statistics, pp. 1–106, Lecture Notes in Mathematics, 1840, Springer, Berlin, 2004.

[Tsi05] Boris Tsirelson. Percolation, boundary, noise: An experiment. Preprint. math/0506269, 2005.

[Wer07] Wendelin Werner. *Lectures on Two-dimensional Critical Percolation*. IAS Park City Graduate Summer School, 2007.

Index

Printed in the United States
by Baker & Taylor Publisher Services